Climate Change and Migration

CLIMATE CHANGE AND MIGRATION

Security and Borders in a Warming World

Gregory White

OXFORD
UNIVERSITY PRESS

Oxford University Press, Inc., publishes works that further
Oxford University's objective of excellence
in research, scholarship, and education.

Oxford   New York
Auckland   Cape Town   Dar es Salaam   Hong Kong   Karachi
Kuala Lumpur   Madrid   Melbourne   Mexico City   Nairobi
New Delhi   Shanghai   Taipei   Toronto

With offices in
Argentina   Austria   Brazil   Chile   Czech Republic   France   Greece
Guatemala   Hungary   Italy   Japan Poland   Portugal   Singapore
South Korea   Switzerland   Thailand   Turkey   Ukraine   Vietnam

Published by Oxford University Press, Inc.
198 Madison Avenue, New York, New York 10016

www.oup.com

Oxford is a registered trademark of Oxford University Press

Library of Congress Cataloging-in-Publication Data
White, Gregory, 1960–
Climate change and migration : security and borders in a warming world / Gregory White.
    p.   cm.
Includes bibliographical references and index.
ISBN 978-0-19-979482-9 (hardback)—ISBN 978-0-19-979483-6 (pbk.) 1. Emigration and immigration—
Social aspects. 2. Emigration and immigration—Environmental aspects. 3. Climatic changes—Social aspects.
4. Global environmental change—Social aspects. 5. Global warming—Social aspects. I. Title.
JV6225.W45   2011
304.8—dc22   2011007330

Printed in the United States of America
on acid-free paper

For Helen Whayne White and George Hamilton White, Jr.

CONTENTS

PREFACE

I have long been deeply interested in migration studies and environmental studies. Yet for many years the two intellectual passions rarely crossed paths. I published and taught courses on the two topics for many years, but basically kept them apart. In 2005, however, the two concerns began to merge, at least in my own thinking. During a seminar on "Green Diplomacy" one of my students presented a superb paper on how environmental change in the Aral Sea Basin had affected the region's population movements. The resultant discussion was stimulating. Yet it was somehow incomplete. We felt comfortable with the discourse of political science: borders, the legacy of Soviet development projects, population displacements, efforts to improve diplomatic cooperation between Kazakhstan and Uzbekistan. However, we were at a loss to engage natural-science issues associated with the ecological devastation in the Aral Sea. Hydrology, the water cycle, soil science, fisheries, and climate science were all clearly relevant, but we had no way of linking those concerns with our focus on political science. How do political factors interact with natural systems and vice versa? We talked at some length about what it would mean to be multi- and interdisciplinary. We imagined what it would be like for scholars to bridge international relations and environmental sciences, for social science to meet natural science. Imagine, I recall suggesting, a political scientist or legal scholar who was trained in environmental science, or vice versa.

That experience pushed me to explore the growing literature on environmental refugees and climate migration—that is, people compelled to travel because of climate change. Climate-induced migration (CIM) has been increasingly invoked in the context of African migration to Europe—a particular interest of mine because of my scholarship on migration politics in North Africa and the Mediterranean Basin. The assertion has been made that immigration to Europe from Africa has and would be further driven by climate change. North African governments have also played a climate migration "diplomatic card" in their interactions with the EU, arguing that immigrants from sub-Saharan Africa were arriving because of environmental stress. Even if the card was not always overtly played, it lurked. The problem is that these assertions were often uncorroborated. Given the ongoing impact of climate change, it seemed reasonable to argue that CIM would grow in the future, but little evidence was offered on the scope of the migration, its direction, or the degree to which it was or would be driven by further climate change. The stakes for both climate change and migration politics are high, so it was striking to me that relatively little had been done to bridge disciplinary boundaries.

I was fortunate, then, to receive an Andrew W. Mellon Foundation New Directions Fellowship. The fellowship gave me the opportunity to begin training in climate science and environmental studies. Columbia University's Earth Institute offered a welcoming milieu. With a leave from Smith College, I was able to enroll in graduate seminars such as "Climate Variability and Climate Change" and "Climate Change in Africa." I attended lectures and symposia at the Earth Institute, the School for International and Public Affairs (SIPA), the Department of Earth and Environmental Sciences, the International Research Institute (IRI) for Climate and Society at the Lamont-Doherty Earth Observatory (LDEO), and the Center for Research on Environmental Decisions (CRED). Faculty, staff, and fellow students at Columbia University were strikingly gracious and generous. Being in New York also allowed me to attend a symposium on climate-induced migration at the United Nations Institute for Training and Research (UNITAR). By no stretch of the imagination can I now claim to be an expert in climate science. That would require a career of scholarship and teaching. Yet I now have a far deeper understanding of and appreciation for the field.

Throughout the experience at Columbia (and my ongoing "retraining") I also valued the opportunity to break down and reexamine my own disciplinary assumptions. At Columbia, I often felt like an ethnographer trying to learn a new language in a distant land populated with natural scientists. I relived long-repressed memories of being a matriculated graduate student standing in line at a beleaguered registrar's office. And I embraced the strangeness I felt as a middle-aged political scientist sitting in an Earth and Environmental Sciences seminar with twentysomething students. They all seemed to understand intuitively the discussion of the dynamic balances in the Earth's atmosphere and that the Coriolis force could be expressed as:

$$(dv/dt)_{Co} = -2\,u\,\sin\emptyset$$

(At least I think that is right.) To put it mildly, I did not understand the equation. I quickly resigned myself to the fact that I would likely *never* understand how one might derive a formula for the conservation of angular momentum. On the other hand, I *could* gain a decent understanding of why the Coriolis force, coupled with friction from the Earth's surface, could prompt a change in the geostrophic wind flow and a convergence of surface winds into centers of low pressure. I began to understand the forces that prompt intertropical convergence zones (ITCZs), Hadley and Walker cells, Ekman transport in ocean gyres, and how the oscillation of warm water and atmosphere in the Pacific Ocean produce the El Niño-Southern Oscillation (ENSO). I was absolutely astonished to find that the ocean surface is sloped. That's right. The slope is only on the order of a few meters over a distance of thousands of kilometers, yet it contributes to the geostrophic currents that, in turn, produce ocean gyres—clockwise in the Northern Hemisphere and counterclockwise in the Southern Hemisphere. The bottom line: despite the fact that I had been "disciplined" as a political scientist for many years, I began to understand the remarkable and stimulating field of earth science. And it helped me to apprehend the natural science basis of climate-induced migration and, especially, the *politics* of CIM.

It was intense and provocative to be conducting my "ethnographic fieldwork" as a student during the run-up to the December 2009 Copenhagen Conference. There was also the ostensible scandal associated with hacked e-mails from the Climate Research Unit at the University of East Anglia, the "revelations" that the Intergovernmental Panel on Climate Change's *Fourth Assessment Report* (AR4) in 2007 contained some errors, and the political controversy over the "Snowmageddon" blizzards of February 2010. Suffice to say, I devoured media coverage associated with the politics of climate science.

My pursuit has also led me to develop a new interdisciplinary course: "Science, Public Policy, and the Politics of Science." As National Public Radio science journalist Ira Flatow has noted, on an annual basis we are treated to news programs and articles on the "Year in Review." The year-end genre is familiar: quick, capsule surveys of world events, electoral politics, celebrity scandals, notable obituaries, and sports highlights. They are interesting and even entertaining. Yet how often are major breakthroughs in scientific knowledge noted? The Large Hadron Collider, the Mars rover, breakthroughs in modeling climate change, in nanotechnology, agronomy, and medicine? When science does make the news, it is often presented as controversial. Should you get mammograms? Should you vaccinate your child? At times it is even cast in faith terms: "I don't *believe* in climate change." The new course examines the role that science plays in our lives as individuals and citizens, the challenges that scientists confront in communicating their findings to the public, and the use of scientific findings by policy makers. It examines critical perspectives on the scientific method and the accumulation of scientific knowledge, and it unpacks the role the media play in the public sphere. I began this preface with an anecdote about teaching. When you teach, you learn. I am forever grateful for the opportunity to teach and learn with talented students at Smith College.

By way of acknowledgments, the Andrew W. Mellon Foundation's New Directions Fellowship provided crucial support for the work that informs the book, especially the research that forms the heart of chapters 2 and 5. Much of that research was conducted while at the University of Columbia. I am especially indebted to Mark Cane and Lisa Goddard of the Department of Earth and Environmental Sciences and the International Research Institute for Climate and Society at the LDEO for their graciousness. Similarly, Alessandra Giannini, also of the IRI, provided valued support and intellectual guidance. Marc Levy, Susanna Adamo, and Alex de Sherbinin at the LDEO's Center for International Earth Science Information Network (CIESIN) offered unstinting encouragement and helped shape the project's focus. I was ever impressed with their kindness and welcoming spirit. Alex was especially helpful.

In Rabat I am indebted to many people, including Mustapha Machrafi of the African Studies Institute (IEA); Mehdi Lahlou of the National Institute of Statistics and Applied Economics (INSEA); and Stéphane Rostiaux, the Rabat bureau chief of the International Organization of Migration. The Jacques Berque Center (CJB) was welcoming. James Miller, Executive Director of the Moroccan-American Commission for Cultural and Educational Exchange also offered support and insight. Rabat-based Fulbright scholars Andrew Waltrous and Kristen Johnson were very helpful, too. As always, Fatema Bellaoui provided friendship and counsel.

In Washington, DC, I benefited from valuable discussions and correspondence with Kathleen Newland of the Migration Policy Institute and Geoffrey Dabelko of the Environmental Change and Security Program of the Woodrow Wilson International Center. I also benefited from feedback and challenging questions from audiences at Princeton University, Williams College, Yale University, the Naval Postgraduate School in Monterey, California, and the World Bank's Center for Mediterranean Integration in Marseille, France. I am especially grateful for the sharp, yet patient challenges offered by Susan Forbes Martin at Georgetown University's Institute for the Study of International Migration in an early presentation to an audience there. Mark Miller of the University of Delaware provided me with intellectual encouragement, as he has done for 25 years.

Smith College also afforded generous assistance, including support for research trips to Tunisia and Morocco. At Smith, I am especially appreciative to colleagues for their support and sharing of ideas, including Alice Hearst, Howard Gold, Patrick Coby, Brent Durbin, and other colleagues in the Government Department; Drew Guswa and colleagues in the Environmental, Science and Policy Program and the new Center for the Environment, Ecological Design and Sustainability (CEEDS); and John Davis, Danielle Smith, Emily Robinson, and David Podboy. Jon Caris and Gretchen Ravenhurst of Smith's Spatial Analysis Lab developed the migrant route map in chapter 4.

In the project's earliest stages, Alan Bloomgarden was a central catalyst. Darcy Buerkle also provided early encouragement. Michael Clancy and Scott Taylor offered guidance. Above all, I am forever grateful to Dana Leibsohn.

The manuscript in various forms benefitted from sharp, critical reading by several people, especially Susan Forbes Martin, Stacy VanDeveer, Scott Taylor, Chantel Pheiffer, and Samantha Riiska. I am also grateful to Dave McBride of OUP for his willingness to hear me out, and to the staff at OUP. Laura MacKay provided expert edits. Of course, with all these words of gratitude, I remain responsible for any errors that remain.

This book is dedicated to my parents, Helen Whayne White and George Hamilton White, Jr. For their part, my children, Sydney and Emmett, are daily reminders of the importance of an "optimism of the will" regarding the future. Finally, Tricia. Her humor and encouragement and love and everything made this book possible.

LIST OF ACRONYMS

AFRICOM	U.S. Africa Command
AFVIC	Friends and Families of Victims of Clandestine Immigration (Amis et familles des victimes de l'immigration clandestine [Rabat])
AMERM	Moroccan Association for the Research and Study of Migration (Association Marocaine d'études et de recherches sur les migrations (Rabat)
AOGCM	Atmospheric-Oceanic Global Coupled Models
AOSIS	Association of Small Island States
AR4	Fourth Assessment Report of the IPCC
ASP	American Security Project
CAN	Center for Naval Analysis (Washington, DC)
CBP	U.S. Customs and Border Protection—part of DHS
CEPOL	European Police College
CEN-SAD	Community of Sahel-Saharan States
CIM	climate-induced migration
CNAS	Center for a New American Security (Washington, DC)
COP15	15th Meeting of Chief of Parties for UNFCCC (2009 Copenhagen Summit)
DHS	U.S. Department of Homeland Security
DOD	U.S. Department of Defense
EEC	European Economic Community
ENSO	El Niño Southern Oscillation
EU	European Union
EMSA	European Maritime Safety Agency
FAO	United Nations Food and Agricultural Organization (Rome)
Frontex	Agency for the Management of Operational Cooperation at the External Borders of the Member States (Warsaw)
G20	Group of 20
GADEM	Anti-Racist Group for the Accompaniment and Defense of Foreigners and Immigrants (Groupe anti-raciste d'accompagnement et de défense des étrangers et migrants [Rabat])
GHG	greenhouse gases
GFMD	Global Forum on Migration & Development
GRACE	Gravity Recovery and Climate Experiment
GWOT	global war on terror
ICE	U.S. Immigration and Customs Enforcement—part of DHS
ICMPD	International Centre for Migration Policy Development (Vienna)

IIRIRA	U.S. 1996 Illegal Immigration Reform and Immigrant Responsibility Act
ILO	International Labor Organization
IOM	International Organization of Migration (Geneva)
IPCC	Intergovernmental Panel on Climate Change
IRCA	U.S. 1986 Immigration Reform and Control Act (the Simpson-Mazzoli Act)
ITCZ	intertropical convergence zone
MAP	Maghreb Arab Press Agency
NAFTA	North American Free Trade Agreement
NAPA	National Adaptation Programmes of Action
NIC	U.S. National Intelligence Council
NIE	National Intelligence Estimate
OMDH	Moroccan Organization for Human Rights (Organisation Marocaine des Droits Humains [Rabat])
QDR	Quadrennial Defense Review
RUSI	Royal United Services Institute (London, UK)
SIDS	Small Island Developing States
SIS	Schengen Information System
SLP	sea level pressure
SST	sea surface temperature
TSCTI	Trans-Sahara Counter Terrorism Initiative
UNCED	UN Conference on Environment and Development (1992 Rio/Earth Summit)
UNEP	UN Environmental Program
UNFCCC	UN Framework Convention on Climate Change
UNHCR	UN High Commissioner for Refugees
WBGU	German Advisory Council on Global Change (Berlin)
WTO	World Trade Organization

Climate Change and Migration

Introduction

Evidence is fast accumulating that, within our children's lifetimes, severe droughts, storms and heat waves caused by climate change could rip apart societies from one side of the planet to the other. Climate stress may well represent a challenge to international security just as dangerous—and more intractable—than the arms race between the United States and the Soviet Union during the cold war or the proliferation of nuclear weapons among rogue states today.
—Thomas Homer-Dixon, 2007[1]

There is a new phenomenon in the global arena: environmental refugees. These are people who can no longer gain a secure livelihood in their homelands because of drought, soil erosion, desertification, deforestation and other environmental problems, together with associated problems of population pressures and profound poverty. In their desperation, these people feel they have no alternative but to seek sanctuary elsewhere, however hazardous the attempt.
—Norman Myers, 2005[2]

Ten adult male migrants from West Africa drowned off the Coast of Gran Canaria Island after their boat capsized, Spanish rescuers say. Five bodies were found floating near a beach and five more were recovered by divers on Friday. Spain's Civil Guard said six other adults survived from the boat, believed to have been carrying at least 18.
—BBC World News, September 7, 2007[3]

In the last decade, North Atlantic security officials became increasingly preoccupied with climate change.[4] This is not surprising. After all, climate change will likely prompt significant geopolitical competition as

1. Thomas Homer-Dixon, "Terror in the Weather Forecast," *New York Times*, April 24, 2007.
2. Norman Myers, "Environmental Refugees: An Emergent Security Issue" (paper presented at the 13th Economic Forum, Prague, Czechoslovakia, 2005).
3. Available at http://news.bbc.co.uk/2/hi/europe/6983158.stm.
4. North Atlantic as a region includes the United States, Canada, the European Union (EU), and its constituent members.

countries endeavor to secure access to oil and natural gas, prevent food shortages, and cope with strategic challenges presented by rising seas and the opening of Arctic sea routes. For example, Diego Garcia, the naval base jointly operated by the United Kingdom and the United States in the middle of the Indian Ocean, is pivotal for military operations in East Africa, the Middle East, and South Asia. Yet this atoll is only 22 feet above sea level at its highest point. As sea levels are anticipated to rise (because of thermal expansion of ocean waters and the melting of terrestrial ice packs), Diego Garcia's ability to sustain a naval air base has been put into question.

Security officials are also devoting more attention to the potential that gradual climate degradation will increase human migration pressures. Warmer temperatures and changing precipitation patterns are expected to accelerate migration and population displacement. Recent analyses of climate-induced migration (CIM)[5] may use different labels to describe it— climate migration, environmental migration, or climate refugees—each with political and analytical effects that are themselves fascinating to examine. But even with the methodological and terminological challenges, the evidence is abundant: a combination of rising sea levels, increasing temperatures, and changing precipitation patterns will likely affect migration patterns in the decades to come.

5. See inter alia Alex de Sherbinin, Koko Warner, and Charles Ehrhart, "Casualties of Climate Change," *Scientific American* 304:1, January 2011, 64–71; Susana B. Adamo, "Addressing Environmentally Induced Population Displacements: A Delicate Task" (Population-Environment Research Network Cyberseminar on "Environmentally Induced Population Displacements," 2008); Koko Warner, Charles Ehrhart, Alex de Sherbinin, and Susana Adamo, *In Search of Shelter: Mapping the Effects of Climate Change on Human Migration and Displacement* (New York: CARE International, 2009); Tina Acketoft, "Environmentally Induced Migration and Displacement: A 21st Century Challenge" (Strasbourg, France: Council of Europe Committee on Migration, Refugees and Population of the Parliamentary Assembly, 2008); Sabine Perch-Nielson et al., "Exploring the Link between Climate Change and Migration," *Climatic Change* 91 (2008), 375–393; Vikram Odedra Kolmannskog, *Future Flood or Refugees: A Comment on Climate Change, Conflict and Forced Migration* (Oslo, Norway: Norwegian Refugee Council, 2008); "Special Issue: Climate Change and Displacement," *Forced Migration Review* 31 (October 2008); Alex de Sherbinin et al., "The Vulnerability of Global Cities to Climate Hazards," *Environment and Urbanization* 19 (2007), 39–64; Dominic Kniveton et al., *Climate Change and Migration: Improving Methodologies to Estimate Flows* (Geneva, Switzerland: International Organization for Migration, 2008); Frank Laczko and Christine Aghazarm, eds., *Migration, Environment and Climate Change: Assessing the Evidence* (Geneva: International Organization for Migration, 2010); and Etienne Piguet, *Climate Change and Forced Migration* (Geneva, Switzerland: UN High Commission for Refugees Policy Development and Evaluation Service, 2008).

This book seeks to examine critically climate-induced migration and to assess its implications for borders, state sovereignty, and security. Rather than resolving the disputes concerning definitions and measurements, the book argues that the disputes themselves merit engagement because of CIM's political implications. Climate security has increasingly joined more familiar rationales for thwarting immigration.

North Atlantic countries have dramatically enhanced border security efforts since the end of the Cold War. Efforts to control immigration have become so thoroughly politicized and so much a part of the electoral landscape that they are almost taken for granted. At times, the incendiary quality of immigration politics are obvious, as in the Arizona Senate's passage of SB 1070 in April 2010. More subtle are debates about public school financing. For decades, such anti-immigrant sentiment generally centered on two rationales: economic and societal security. Immigrants were cast not only as threats to jobs but also as different socially and therefore hard to integrate. This is not to suggest that such concerns were entirely unfounded, or that there were not real issues concerning assimilation or integration. Yet many opponents of immigration underestimated immigrants' contributions to host countries and, especially in the United States, their own families' immigration experiences. It is easier, too often, to emphasize that immigrants are catastrophically different.

More recently, beginning in the '80s and '90s, a third rationale—national security—emerged as an anti-immigrant justification.[6] The reasoning shifts depending on the context and the argument, but admixing the three rationales together one might hear, "Not only do immigrants take our jobs, *they* are different from *us*. And they have terrorist or criminal intentions."

Over the last five years or so, something new has happened. In some ways, the shift has been subtle, but *climate security* is increasingly invoked as a fourth justification for robust interdiction efforts by militaries and border patrols. The concern is that "environmental refugees" will add to the existing flow of migrants. Those who analyze the impact of environmental factors on human migration are not always thinking in terms of securing borders. Yet officials charged with anticipating threats to security are. For example, the U.S. National Intelligence Council issued a National Intelligence Estimate in June 2008 on the national security implications of climate change that highlighted the anticipated threat of CIM to the United

---

6. See Christopher Rudolph, "Security and the Political Economy of International Migration," *American Political Science Review* 97: 4 (2003), 603–620.

States and its allies.[7] Similarly, in 2009 U.S. Senators John Kerry (D-MA) and Barbara Boxer (D-CA) introduced S 1733, titled "Clean Energy Jobs and American Power Act." The bill directly invoked CIM as a security concern.[8] Finally, the 2008 National Defense Authorization Act required the U.S. Department of Defense to include climate change in its 2010 Quadrennial Defense Review (QDR). The QDR, released in February 2010, emphasized CIM as a security threat.[9] For their part, European officials regularly invoke climate change as a "threat multiplier" that will augment migration to Europe from the south and east.

This preoccupation with preparing for a potential increase in climate-induced migration seems not to include a concern for ethical responsibility. In *The Ethics of Territorial Borders*, John Williams examines the constructed nature of the Westphalian state system.[10] Williams works in the English School tradition, a theoretical approach within the field of international relations that examines evolving norms and practices in "international society."[11] He argues that borders are artifices and constructs—the product of shared practices between actors in the international arena. As semiotic deployments, borders send messages not only to internal audiences such as citizens but also to external actors such as diplomats and, indeed, would-be migrants and refugees. These constructed and symbolic dynamics can, depending on the circumstance, become hardened with fencing, razor wire, motion detectors, and patrols.

Yet as insightful as Williams's examination is, like many similar treatments in international relations theory it does not consider migration and refugee issues.[12] In contrast, this book focuses on the ethical dimensions

7. Permanent Select Committee on Intelligence and House Select Committee on Energy Independence and Global Warming, "Testimony by Thomas Fingar, Deputy Director for National Intelligence, on the National Security Implications of Global Climate Change to 2030," Washington, DC: U.S. House of Representatives, June 25, 2008.

8. Text available at the U.S. Senate Environment and Public Works Committee, epw.senate.gov.

9. Herbert E. Carmen, Christine Parthemore, and Will Rogers, *Broadening Horizons: Climate Change and the US Armed Forces* (Washington, DC: Center for a New American Security, 2010).

10. John Williams, *The Ethics of Territorial Borders: Drawing Lines in the Shifting Sands* (London: Palgrave, 2006).

11. Andrew Linklater and Hidemi Suganami, *The English School of International Relations: A Contemporary Reassessment* (London: Cambridge University Press, 2005); and Hedley Bull, *The Anarchical Society: A Study of Order in World Politics* (New York: Columbia University Press, 1977).

12. For a superb review of the political science literature and its relative neglect of immigration, see James F. Hollifield, "The Politics of International Migration: How Can We 'Bring the State Back In?'" in *Migration Theory: Talking Across Disciplines*, eds. C. B. Brettell and J. F. Hollifield, 2nd ed. (New York: Routledge, 2007), 183–237.

of climate change and the enforcement of borders.[13] As industrialized countries contribute the most to climate change through consumption and emissions, CIM constitutes an ethical dilemma that will require them to reconsider and revise the existing dialogue concerning migration. Toward that end, this book argues against a security-minded response to CIM.

"Getting tough"—responding in a militarized fashion—is an easy, cynical step in a warming world. It may be politically successful with anxious electorates. It may tap into the public's fears about climate change and the prospect of desperate hordes of "refugees" inundating North Atlantic borders. And it may be more politically palatable than policies that mitigate greenhouse gas (GHG) emissions. Building a fence is easier than changing lifestyles. Yet the injection of security imperatives into climate-induced migration is unethical and unworkable.

It is also an overreaction, because the scientific evidence is persuasive that migration patterns, to the extent that they are attributable to anthropogenic climate change, are *not* changing toward an inundation of North Atlantic borders. People migrate for complicated reasons. But where climate change is prompting increased migration—primarily in the tropical regions of the world—most of that movement is over a short distance, within a subregion. Climate change has and will enhance the vulnerability of populations, yet it is not clear it will induce long-range migrations.

Climate change is happening and demands attention. Policy makers, citizens, and scholars need to treat it directly. People around the world need to mitigate the emissions of greenhouse gases and pursue adaptation. Yet stoking fears of impending streams of climate refugees and militarizing borders are political acts that must be avoided. Among other things, the securitization of CIM sets in motion a "CIM security dilemma." The ostensible securitization of each border merely reassigns responsibility for contending with mixed migration to adjacent or even far-flung borders. The result is a successive transfer of obligation to another part of a border, another country, even another region of the globe.

This brings into view another implication of CIM: the playing out of the security dilemma in "transit states" such as Mexico, Morocco, Tunisia, Libya, Turkey, and others. It is in these "borderlands," through which migrants travel in their efforts to reach North Atlantic countries, that the United States and the European Union exert power. In fact, North Atlantic

13. Stephen Gardiner, "A Perfect Moral Storm: Climate Change, Intergenerational Ethics, and the Problem of Corruption," in *Political Theory and Global Climate Change*, ed. Steve Vanderheiden (Cambridge, MA: MIT Press, 2008), 25–42. For Gardiner, corruption is not "merely" financial, although that is part of the picture; it is also moral.

countries are keen on externalizing their border controls and pressuring their neighbors to help stop the flow. The goal is a kind of "remote migration control" to interdict flows before they ever reach North Atlantic borders.[14] In this sense, sovereignty is moving away from the territorial borders of North Atlantic countries and extends deep into adjacent transit states. For the transit state, this thickening of borders results in complicated dynamics internally, as people who travel into it can go no farther. Some transit states bemoan their role as a policeman for rich countries. More commonly, though, transit states embrace the task, since it can serve a government's geopolitical ambitions. It can even facilitate a "rebordering" of the country.[15] After Libya's renunciation of weapons of mass destruction in 2003, Muammar Qaddafi's regime parlayed North Atlantic thirst for oil and Libyan willingness to interdict transit migrants into concessions on trade and increased financial aid.[16] Mexico's forceful efforts on its southern border with Guatemala and Belize have been undertaken in the context of its central role in NAFTA.[17] And as examined in chapter 4, King Mohammed VI of Morocco has publicly invoked anxieties about CIM and trumpeted his country's efforts in stopping African migration to Europe.

For the transit state, playing the climate-refugee card is not only an effort to enhance its diplomatic status but also part of a transformation of governance that concerns policy, state-society relations, and claims over disputed territory. By exploiting a "threat-defense" logic vis-à-vis CIM, transit states engage in acts of state building, transform state institutions, participate in geopolitical frameworks, and assert sovereignty over borders that are often contested. The cruel irony is that transit states' treatment of aspiring immigrants and transit migrants in their own countries can be harsh and forbidding—even as they seek protected and enhanced status for their emigrants living and working in North Atlantic countries.

These issues are a matter of life and death. Loss of life stemming from efforts to thwart access to North Atlantic countries is daily and inestimable. Conventional news accounts of people dying in shipping containers, in

14. Gerald Kernerman, "Refugee Interdiction before Heaven's Gate," *Government and Opposition* 43: 2 (2008), 230–248.

15. Peter Andreas, *Border Games: Policing the US-Mexico Divide* (Ithaca, NY: Cornell University Press, 2000).

16. Derek Lutterbeck, "Migrants, Weapons and Oil: Europe and Libya After the Sanctions," *Journal of North African Studies* 14: 2 (2009), 169–184.

17. Clyde Hufbauer and Gustavo Vega-Cánovas, "Whither NAFTA: A Common Frontier?" in *The Rebordering of North America: Integration and Exclusion in a New Security Context*, eds. Peter Andreas and Thomas Biersteker (New York: Routledge, 2003), 128–152.

deserts, or, as the quote from the BBC World Service shows, in boat interdictions off Spain's coasts are often vague and callous. They occlude not only geopolitics but also individual tragedies. According to the BBC, for example, "30,000" migrants were caught in 2006 trying to cross from Morocco and the Western Sahara to the Canary Islands. Those who have perished in attempts to cross the Mediterranean or the Arizona border or the Saharan desert are often presented in round numbers, too. Alas, history counts its dead in round numbers. José Seguro, a member of the government in the Canaries, lamented the number of deaths: "I would dare to say that in recent months we have lost hundreds of anonymous migrants."[18] Migrants often lack papers. If their bodies are found, it is true: they are anonymous. It is virtually impossible to identify their country of origin. One wonders if this is an extreme example of what Giorgio Agamben refers to as "bare life," or life existing outside the boundaries of a sovereign, political order.[19]

Migrants who do move across international borders are prompted by a wide array of factors, including slow-onset environmental degradation. Citizens and policy makers need to mitigate GHG and aerosol emissions. Adaptive strategies need to be devised. And encouraging sound development policies—and support and relief to people who are displaced because of environmental degradation—is an ongoing challenge for the international community. Nevertheless, a securitized response to CIM is ultimately counterproductive. It is a politically expedient, short-term maneuver that will only exacerbate the underlying circumstances, not solve the challenges posed by climate change and population movements. As CIM and environmental security have joined with other rationales, officials in Vienna, Rome, Madrid, London, Brussels, and Washington are working more closely with their counterparts in Tripoli, Rabat, Mexico City, and Istanbul. Thus, CIM is becoming an integral component of the ongoing elaboration of a transnational security state devoted to migration controls, along with counterterrorism and drug interdiction. This has crucial implications for North Atlantic citizens, of course, not to mention for people in tropical regions dealing with the challenges of climate change.

This book is organized as follows: chapter 1 introduces climate-induced migration as an "essentially contested concept." It notes the definitional

18. "El Delegado en Canarias denuncia la muerte 'anónima' de cientos de personas en el mar," *El País*, March 7, 2006.
19. Georgio Agamben, *Homo Sacer: Sovereign Power and Bare Life*, trans. Daniel Heller-Roazen (Stanford, CA: Stanford University Press, 1998).

challenges and the evolution of a highly dynamic literature since CIM first emerged as a concept in the mid-'80s. It offers a typology that specifies different kinds of population movements and explores the different dimensions of the debate. It seeks to tread the complicated middle ground between alarmist anticipation of multitudes of desperate refugees at one extreme and dismissive criticisms of the concept on the other.

Chapter 2 explores the empirical phenomenon itself. Humans have always migrated in response to a complex array of stimuli and "forcings." Intruders into the Roman Empire were often motivated by crop failure and famine, and Roman centurions discouraged entry in spectacular and grisly fashion. While Oklahomans flight from the dust bowl in the '30s was spurred by catastrophic policy decisions and sheer venality, Steinbeck's "grapes of wrath" fermented in the context of climate variability that induced a forbidding drought and attendant soil erosion. Despite the historical precedents, chapter 2 takes seriously the natural scientific evidence that climate change has accelerated in the twentieth century because of anthropocentric contributions and that climate change will deepen further in the twenty-first century. As a result, migration patterns are likely to change profoundly.

CIM is a worldwide phenomenon and obviously an issue for Small Island Developing States (SIDS) confronting rising sea levels. (SIDS were first recognized as a diplomatic entity at the 1992 United Nations Conference on Environment and Development in Rio de Janeiro.) CIM is also salient in South Asia, especially in Bangladesh on an annual basis and, for example, in Pakistan in the aftermath of the August 2010 floods. Nonetheless, this book drills into the geographical space associated with African migration to Europe and its implications for governance and transit states. Much of the population movement ostensibly directed toward Europe emerges from the Sahel and sub-Saharan Africa. The Sahel is the zone of transition between the Sahara desert to the north and the savannas and tropical jungles of equatorial Africa to the south; it includes portions of Senegal, Mauritania, Mali, Burkina Faso, Chad, Niger, Nigeria, and the Sudan. Changing temperatures and precipitation patterns are likely to affect migratory pressures in the region. And since Africa is geographically proximate to Europe, it is the most immediate concern for North Atlantic interests. Importantly, however, chapter 2 emphasizes that while climate change may continue to contribute to CIM, most migrants within the region move short distances because adverse environmental conditions *reduce* access to the resources they need to migrate. The research on migration in the Sahelian and sub-Saharan context is that, paradoxically, climate change may inhibit long-range migration. Moving from Niger south to Kano might be enough for an individual to encounter alternative livelihoods, although the experience is surely challenging. So while CIM pressures to the Mediterranean and Europe are hardly insignificant, the bulk of CIM's impact has been, and will likely remain, felt south of the Sahara—*not* on North Atlantic borders.

This fact provides the context for chapter 3 and its examination of security's place at the heart of international relations theory. The chapter also situates CIM at the nexus of complementary discourses. On the one hand, recent decades have seen a growing emphasis on the securitization of the environment and climate change. Concomitantly, on the other hand, there has been a gradual injection of security imperatives into the interdiction of immigration. Climate-induced migration bridges and knits together these discourses to provide a new, additional rationale for security officials and electorates anxious about immigration and the impact of climate change. Since most CIM is not directed toward the North Atlantic, the political reaction serves an ancillary purpose as part of anti-immigration efforts.

Chapter 4 turns to North Africa, known in Arabic as the Maghreb. The chapter focuses on Morocco as a way of illuminating the role of transit states situated "in-between" sending and receiving dynamics. Admittedly, "transit state" is a bit of a misnomer, as migrants are more often blocked and not really in transit. Nonetheless, the label "host country" or "country of immigration" does not work either; the new population does not comprise immigrants who are seeking to settle, as is the case in advanced-industrialized economies. Bearing in mind the labeling issues, chapter 4 treats the politics of CIM *within* a transit state and the ways in which CIM is used to "reborder" a country, cement territorial claims, and control the national space. CIM is also used by transit states as a bargaining chip to enhance the status of their own emigrants—both legal and undocumented—living in North Atlantic countries. Finally, chapter 4 treats the ways in which CIM enhances collaboration between North Atlantic and transit state officials and facilitates the elaboration of a transnational security state—that is, the internationalization of security apparatuses and interior ministries.

Chapter 5, in conclusion, examines the challenges of incorporating CIM into international relations theory. Although the limitations and pitfalls of a security-minded approach are evident, the challenges of "desecuritizing" climate-induced migration are profound. Is it enough to merely point out the securitization? What then? Entrenched interests are keen on sustaining militarized borders, detention centers, and lucrative smuggling networks. Fences make good politics. Good business, too. Security measures bolster the profits of traffickers, which in turn justify further security measures. Moreover, the international community is not in a position to extend recognition to—or contend with the challenges posed by—climate refugees. It is difficult enough to protect "convention refugees," the political refugees protected under the 1951 United Nations Convention Relating to the Status of Refugees.

Nonetheless, chapter 5 examines the opportunities for enhanced cooperation, especially at the regional level. It seeks to contribute to efforts to steer the normative discourse away from security and toward a more nuanced and constructive approach to climate-induced migration, one that emphasizes improved governance and a focus on "development and climate" initiatives. Emphasizing that CIM is a real phenomenon, yet one that is not expected to induce a rush toward North Atlantic borders, the book concludes with a call for deeper efforts to mitigate GHG emissions, development policies that promote genuine sustainability, and a reduction in support for security initiatives and military expenditures by North Atlantic and transit states that do not address the challenges posed by climate change. This is a tall order. Nevertheless, it is preferable to a misguided security orientation. In the context of climate change, "adaptation" has long implied resigned acceptance and/or giving up on mitigation. Yet it need not be cast as acquiescence. Adaptation is a dynamic, innovative, and creative process that must be pursued.

CHAPTER 1

Climate-Induced Migration

An Essentially Contested Concept

Is climate change forcing people to travel across international borders in unprecedented numbers? If yes, should "climate migrants," or "environmental refugees," or "climate refugees," be understood differently than economic migrants or political refugees? Should they be permitted to cross a secure border and offered protection? Or should they be apprehended, detained, and repatriated to their country of origin? Finally, should fences be built and fortified to stop unwanted immigration? These questions form the heart of this book. If there were no such thing as climate-induced migration (CIM), then there would be nothing to contemplate. Yet climate change has occurred—and is likely a growing contributor to transborder movements—so international society's response becomes a matter of deep concern. CIM implicates the wide array of relationships of authority between individuals, civil society, and governments. It also complicates diplomatic and political relations between states. Moreover, if climate change is forcing increasing numbers of people to try to cross international borders in search of safety and refuge, then CIM is an ethical concern of the first order.

In the post—Cold War era, as understandings about the causes and impacts of climate change have deepened, states have become preoccupied with border security. Not because of climate change, of course; it is, quite literally, coincidental. The politics of border security is more directly linked to the dynamics of globalization.[1] Today, "globalization" is conventionally

---

1. See inter alia Peter Andreas, "Redrawing the Line: Borders and Security in the Twenty-First Century," *International Security* 28: 2 (2003), 78–111; Thomas Friedman, *The Lexus and the Olive Tree* (New York: Farrar, Straus, & Giroux, 1999); Saskia Sassen, *Losing Control?*

seen as a phenomenon tied to the end of the fall of the Berlin Wall in 1989 and the collapse of the Soviet Union. The end of Cold War competition was certainly crucial to the emergence of globalization. One might argue, instead, that the '70s are the period of genesis for post—Cold War globalization. Events such as the removal of the U.S. dollar from the gold standard in 1971, oil shocks engineered by OPEC in 1973 and 1979, the emergence of microchip technology, the deepening integration of the European Economic Community, and China's transition in the aftermath of Mao's death in 1976 are central to understanding globalization in the '90s and '00s.

Whatever its causes and origins, even as the international arena has become more open to flows associated with finance, trade, communication, and culture, countries have tightened their physical borders. At the annual World Economic Forum in Davos, Switzerland, bankers "boast" about a borderless world.[2] For this era's international migrants, however, international borders have become intensified. In fact, in a rapidly globalizing world, countries *can* assert sovereignty around migration control, even as their ability to do so in other realms may be constrained.[3]

In particular, North Atlantic countries have deepened border controls by building fences, deploying patrols, and installing state-of-the-art surveillance technologies. For example, in 1994, the United States commenced Operation Gatekeeper along the Mexican border, with efforts to militarize border control continuing apace ever since. Building and securing walls along the U.S.-Mexican border is a perennial issue at the federal, state, and local levels. Immigration politics play out in Europe, too. In the '80s, the European Union (EU) began to establish the Schengen Information System (SIS), a passport control regime that now includes most of its members. In 2005, the EU went further, establishing the Agency

Sovereignty in an Age of Globalization (New York: Columbia University Press, 1996); and Joseph Stiglitz, *Globalization and Its Discontents* (New York: Penguin Putnam, 2002).

2. For a useful distinction between different views of globalization—the "banker's boast," the "social democrat's lament" and the "postmodern dance"—see Frederick Cooper, *Colonialism in Question: Theory, Knowledge, History* (Berkeley: University of California Press, 2005).

3. Matthew Sparke, "Political Geography: Political Geographies of Globalization (1)—Dominance," *Progress in Human Geography* 28: 6 (2004), 777–794; Virginie Guiraudon and Gallya Lahav, "A Reappraisal of the State Sovereignty Debate: The Case of Migration Control," *Comparative Political Studies* 33: 2 (2000), 163–195; Ken Conca, "Rethinking the Ecology-Sovereignty Debate," *Millennium* 23: 3 (1994), 701–711; Stephen Krasner, *Sovereignty: Organized Hypocrisy* (Princeton, NJ: Princeton University Press, 1999); Robert Jackson and Carl Rosberg, "Why Africa's Weak States Persist: The Empirical and the Juridical in Statehood," *World Politics* 35: 1 (1982), 3–32; and Michael Barnett, "The New United Nations Politics of Peace: From Juridical Sovereignty to Empirical Sovereignty," *Global Governance* 1: 1 (1995), 45–61.

for the Management of Operational Cooperation at the External Borders of the Member States. Conveniently known by the shorter name, Frontex, the Warsaw-based agency is responsible for coordinating the policing of Europe's external borders.

Countries also expend great effort on the semiotics of vigorous border security. The goal is to send a symbolic message to aspiring migrants that they are not welcome. In 2005, for example, the U.S. Border Patrol spent $1.5 million for airtime on Mexican radio and TV for "No Mas Cruces en La Frontera"—"No More Crucifixes on the Border"—a campaign to discourage immigrants.[4] Similarly, in 2007 Spain broadcast spots across West Africa television showing bodies washed up on Spanish shores and sobbing mothers who had not heard from their emigrating children. The legendary Senegalese singer Youssou N'Dour intones at the end, "You already know how this story ends. Thousands of destroyed families. Don't risk your life for nothing. You are the future of Africa."[5]

Of course, the semiotics of border control also play—perhaps even more so—to domestic audiences anxious about "societal security" and the ostensible invasion of immigrants.[6] For advanced-industrialized societies, immigrants and the immigration process have long posed both opportunities and challenges—in the workplace, in assimilation and integration dynamics, in culture and the arts. It would not do to suggest that all concerns about the impact of immigration are unfounded. Yet there is no denying that some politicians and media personalities stake their careers on anti-immigrant positions, stoking debates with fear tactics and gaining traction from being "tough" on immigration. The bottom line is that in recent decades, immigration has become thoroughly "electoralized" in the North Atlantic.

In the aftermath of World War II, advanced-industrialized countries relied on immigrant labor to power rapidly recovering economies. Well into the '70s, a migrant could cross the Mediterranean and enter Europe with relative ease. The border between Mexico and the United States was fairly permeable, too. It would be absurd to suggest that it was a simple journey, or that there was no control, or that life was easy as soon as one arrived. Nonetheless, the contrast to today is striking. Governments have demonstrably militarized borders. Radar, motion and acoustic detection

---

4. Richard Marosi, "Border Patrol Tries New Tune to Deter Crossers," *Los Angeles Times*, July 4, 2005.

5. Video clips are available at http://news.bbc.co.uk/2/hi/africa/7004139.stm.

6. See Ole Wæver, "Societal Security: The Concept," in *Identity, Migration and the New Security Agenda in Europe*, eds. Ole Wæver et al. (New York: St. Martin's Press, 1993), 17–40.

systems, razor wire, double fences, and airborne and seaborne patrols are commonplace. New efforts to patrol and thwart unwanted immigration are constant. Whether these measures work or are in fact counterproductive is an important issue.[7] Above all, however, the choice to securitize borders illuminates them as ethical sites and raises deep concerns about the nature of justice. Is it fair that some are privileged by chance to be born in one space while others are shut out? If some are allowed in, on what criteria? Who decides who is allowed to enter? Why are others rejected?

Moreover, assertions of control illuminate the fact that borders are the product of dynamic historical forces. A familiar political map might have green, pink, and yellow countries with borders rendered as thin, two-dimensional lines. Weather reports on any country's evening news invariably show a clearly outlined national space. In the United States, weathercasters often display a map of temperatures and pressure fronts in which the country is suspended in space, with no depiction of Canada or Mexico. Canada only gets mentioned when it is to blame for a cold front. On Moroccan television, weather reports similarly further a nationalist agenda. Maps show no line demarcating the disputed Western Sahara territory, an important symbolic assertion of Morocco's territorial claim to control. Neighboring Algeria or Spain might be depicted, but usually not.

Such two-dimensional conceptions are inadequate. We know that borders exhibit third and fourth dimensions. In the third, spatial dimension, borders have varying degrees of "thickness": in many respects, borders can extend far beyond the official frontier. Technological innovations have deepened sensory capacities and the ability to detect intruders. A migrant might not even contemplate heading for a border hundreds of miles away because of received wisdom about the difficulties of crossing. She may be intimidated by broadcasted images on Mexican or Malian television. Or she may be in contact by cell phone or the Internet with relatives and communities already at the destination. In either case, she is experiencing the border many miles before reaching it. If she is able to cross, she may continue to experience the border thousands of miles beyond the line on the map. A workplace raid in Gary, Indiana, is effectively an exertion of border control. Arizona's SB 1070 undermines integration and assimilation processes by compelling a migrant to remain deeply ensconced within immigrant communities, rather than venture into areas where she would be racially and linguistically profiled.

---

7. Randal Archibold, "As Mexico Border Tightens, Smugglers Take to Sea," *New York Times*, July 18, 2009.

As for the fourth dimension, time, borders are highly mutable. They emerge in history's contests, the give-and-take between countries, and the sovereignty bargains made between political actors. A few borders, being time-honored and geographically demarcated, are relatively clear. A river, a body of water, or a mountain range can do wonders for a border. The Strait of Gibraltar clearly constitutes a boundary between modern Morocco and Spain. (Actually, even that has been challenged by Osama bin Laden's call for the reunion of al-Andalus and the reestablishment of an Islamic Caliphate.) Most borders, however, are contingent and the product of artifice. We can study their histories and their ostensible certainties, often dating their establishment. We can also assess their porosity. As borders are thickened or thinned, as they evolve, and as choices are made to securitize them, their ethical dimension becomes even clearer.

And what of climate change? The likelihood that people are motivated to move because of climate change further complicates the ethics of border security. Migrants have been conventionally divided into those who are "pushed" because of economic calculations and those who move because of political factors. Of course, the distinction in motivation at the individual, micro level can be very murky.[8] Moreover, the attempt to distinguish between an economic migrant and a political refugee often implicates broader historical dynamics and contexts. Why is a Dominican arriving in Florida an economic migrant, while a Cuban is a political refugee worthy of legal protection? Considering climate change as an additional push factor muddies the issues still further.

Since the '80s, the degree to which climate change and environmental change is forcing people to migrate has been receiving more and more attention in the scholarly literature. Some analyses offer mind-boggling estimates of the volume of people expected to move. Others offer skeptical assessments of climate change's actual impact on migration. The latter perspective is complicated. Some in this skeptical camp are suitably doubtful, noting the challenges of ascertaining the causes of migration. Others are wary of the political uses of the concept of climate refugees. Still a third group questions whether climate change is even happening; if it is, they argue, it is not anthropogenic but the product of natural, geological transformation.

8. Stephen Castles and Mark Miller, *The Age of Migration: International Population Movements in the Modern World*, 4th ed. (New York: Palgrave Macmillan, 2010); Douglas Massey et al., "Theories of International Migration: A Review and Appraisal," *Population and Development Review* 19: 3 (1993), 431–466; G. J. Borjas, "Economic Theory and International Migration," *International Migration Review* 23: 3 (1989); Michael Todaro, *International Migration in Developing Countries: A Review of Theory* (Geneva, Switzerland: International Labour Organization, 1976).

Some go so far as to argue that the whole notion of anthropogenic climate change is an elaborate hoax designed to scare people and harm advanced-capitalist economies. In his 2004 novel *State of Fear*, the late Michael Crichton suggested that environmental activists engage in terrorist activities in order to cause natural calamities and the deaths of innocent victims.[9] The ultimate goal of environmentalists, according to Crichton, is to induce a "state of fear" in the polity, thereby empowering the government to cow the governed into submissiveness. It would be easy to dismiss the book as a laughable, poorly written yarn—suitable at best for a summer beach read—were it not for the fact that after its publication U.S. Senator James Inhofe (R-OK) invited Crichton to testify before the Senate Committee on Environment and Public Works. President George Bush hosted Crichton at the White House. And conservative pundits like Rush Limbaugh, Sean Hannity, and Glenn Beck interviewed and sang praise songs to Crichton. For his part, Limbaugh used Crichton-style reasoning on April 29, 2010, in theorizing that environmentalists and the Obama administration conspired to blow up British Petroleum's Deepwater Horizon oil rig in the Gulf of Mexico to undermine efforts to expand offshore drilling. Seriously.[10]

Although it is important to engage self-styled "environmental skeptics" or climate change deniers, the climate change debate is not of interest here. As chapter 2 makes clear, the evidence is persuasive that climate change is occurring and that human activity is contributing to it in complicated, multifaceted ways. Moreover, the globe will continue to warm in coming decades, primarily because of the climate system's lag in responding to "forcings" such as human-made aerosols and greenhouse gases—carbon dioxide (CO_2), methane (CH_4), nitrous oxide (N_2O), and chlorofluorocarbons. As climate scientists argue, ample thermal inertia is already in place— "global warming in the pipeline"—due to the Earth's current radiation imbalance.[11] The precise character of future climate change may remain unknown, but state-of-the art modeling and scenario building demonstrate that climate change will continue apace, if not accelerate.[12]

9. Michael Crichton, *State of Fear* (New York: Avon Books, 2004).

10. See www.rushlimbaugh.com/content/home/daily/site_042910/content/01125113.guest.html.

11. James Hansen et al., "Earth's Energy Imbalance: Confirmation and Implications," *Science* 308 (June 3, 2005), 1431–1435.

12. Gavin A. Schmidt, "The Physics of Climate Modeling," *Physics Today* (January 2007), 72–73; Meinrat O. Andreae et al., "Strong Present-Day Aerosol Cooling Implies a Hot Future," *Nature* 435: 30 (2005), 1187–1191; Alexandra Witze, "Losing Greenland," *Nature* 452 (April 2008), 798–802; and L. R. Kump et al., *The Earth System* (Upper Saddle River, NJ: Prentice-Hall, 2004).

The responsibility for climate change, then, becomes an issue in the ethical consideration of border security. The evidence is compelling that climate change is a cause of international migration flows. It is rarely, if ever, the only cause, yet it is a recursive causal factor that needs to be understood in a sophisticated fashion. Accentuating the moral and ethical issues associated with border security is advanced-industrialized countries' outsized contribution to climate change. Is it ethical for people moving because of ecological changes wrought by industrialization to be barred from the spaces of the people who caused the problem? To what extent are advanced-industrialized countries implicated in the straitened circumstances of poor countries? What is the responsibility of current generations to future generations—namely "intergenerational justice"?[13] How does one think about the growing contribution to climate change of rapidly industrializing countries such as China and India? Until 2008, the United States was the largest emitter of CO_2 and other greenhouse gases; China now leads. Yes, the United States remains the largest emitter on a per capita basis. Yet given the size of China's population and growing economy, China is likely to remain foremost in total emissions. Further, although the preoccupation here is with international borders, internally displaced peoples (IDPs) within countries as large as China, India, or Sudan meet security measures when they migrate. At bottom, therefore, there is a profound need to reconsider the security discourse concerning CIM. This requires a careful analysis of the phenomenon itself, as well as a consideration of the potential responses on the part of the international community.

Before turning to the CIM's emergence as a contested concept and the development of a typology of migrant motivations and political positions, it is important to be ever mindful of the human dimension. George Orwell's passionate 1946 essay "Politics and the English Language" instructs that a concept's legacy is essential. One must also bear in mind that political analyses can become anodyne. Orwell writes:

> In our time, political speech and writing are largely the defense of the indefensible. Things like the continuance of British rule in India, the Russian purges and deportations, the dropping of the atom bombs on Japan, can indeed be defended, but only by arguments which are too brutal for most people to face,

13. Burns Weston and Tracy Bach, "Recalibrating the Law of Humans with the Laws of Nature: Climate Change, Human Rights, and Intergenerational Justice" (Vermont Law School Legal Studies Research Paper Series No. 10–06, 2009); and Gardiner, "A Perfect Moral Storm: Climate Change, Intergenerational Ethics, and the Problem of Corruption," in *Political Theory and Global Climate Change*, ed. Steve Vanderheiden (Cambridge, MA: MIT Press, 2008), 25–42.

and which do not square with the professed aims of the political parties. Thus political language has to consist largely of euphemism, question-begging and sheer cloudy vagueness. Defenseless villages are bombarded from the air, the inhabitants driven out into the countryside, the cattle machine-gunned, the huts set on fire with incendiary bullets: this is called *pacification*. Millions of peasants are robbed of their farms and sent trudging along the roads with no more than they can carry: this is called *transfer of population* or *rectification of frontiers*. People are imprisoned for years without trial, or shot in the back of the neck or sent to die of scurvy in Arctic lumber camps: this is called *elimination of unreliable elements*. Such phraseology is needed if one wants to name things without calling up mental pictures of them.[14]

Thus, while one needs to drive to the heart of CIM as a crucial issue and theoretical construct with important implications for policy and theory, the real-world, human implications of migration remain central. The exploration of categories such as climate refugees or climate-induced displacement cannot slip into a "defense of the indefensible." Climate refugees pose an ethical, humanitarian question, and a "mental picture" and empirical analyses of vulnerable individuals and families struggling in a difficult world should not be lost in theoretical or policy debates.

AN ESSENTIALLY CONTESTED CONCEPT

Climate-induced migration is a broad, disputatious category that includes "environmental refugees" and "climate migrants." Some have even called the phenomenon "climigration."[15] The label one uses has a profound impact on the images and messages conveyed. Suffice to say, the definitional clash is robust. Grappling with different labels and meanings is crucial. Upon first contact, one might have an ostensible understanding of what a given term means. Yet, as the concept is unpacked, and as different analysts deploy different labels with different methodologies and different normative orientations, the disputes become profound. As argued throughout this book, and elucidated further in chapter 3, this lack of definition and fundamental inability to agree about a concept provides the political opportunity for the securitization of CIM and a deeper obsession with border security.

14. George Orwell, "Politics and the English Language," in *A Collection of Essays* (Orlando, FL: Harcourt, 1970).
15. Robin Bronen, "Forced Migration of Alaskan Indigenous Communities Due to Climate Change: Creating a Human Rights Response" (Fairbanks, AK: University of Alaska Resilience and Adaptation Program, 2008).

By way of backdrop, the notion of environmental refugees is conventionally cited as having first appeared in a 1985 United Nations Environmental Program paper. In it, El-Hinnawi defines environmental refugees as individuals who are "forced to leave their traditional habitat, temporarily or permanently, because of a marked environmental disruption (natural and/or triggered by people) that jeopardized their existence and/or seriously affected the quality of their life."[16] Kibreab points out, however, that the concept appeared for the first time in a 1984 International Institute for Environment Development paper. Regardless, Kibreab further argues that by the '90s environmental refugees had become a felicitous "catch-all term."[17]

Initially, the overriding assertion was that environmental degradation was forcing population displacement that would, in turn, prompt refugee crises for advanced-industrialized countries.[18] Writing at the end of the Cold War, Kaplan prophesized "the coming anarchy" and the strain that hordes of migrants would pose to the infrastructure of northern countries, to their capacity for absorption, and to their ability to contend with cultural differences.[19] Consistent with Huntington's 1993 "clash of civilizations" thesis, Kaplan posited that Western culture would be "weakened by cultural disputes."[20] Such Yeatsian arguments envisioned a "mere anarchy loosed upon the world" and dovetailed neatly with film portrayals at the time, too. *The March*, David Wheatley's sensationalist feature-length film, which appeared on the BBC in 1990, depicted multitudes of Africans streaming toward Europe. Xavier Koller's *Journey of Hope* (1991) and Gregory Nava's *El Norte* (1983) were more micro-oriented in their accounts of individuals or families seeking to enter Europe and the United States. Yet the films had the same message: there are people in the "third world" who will risk everything, including their lives and the lives of their children, to get to affluent countries.[21]

16. Essam El-Hinnawi, *Environmental Refugees* (Nairobi, Kenya: United Nations Environment Program, 1985).

17. Gaim Kibreab, "Environmental Causes and Consequences of Migration: A Search for the Meaning of 'Environmental Refugee,'" *Disasters* 21: 1 (1997), 20–38.

18. Thomas Homer-Dixon, "On the Threshold: Environmental Changes as Causes of Acute Conflict," *International Security* 16: 2 (1991), 76–116; and Norman Myers, "Environmental Refugees," *Population and Environment* 19 (1997), 167–182.

19. Robert Kaplan, "The Coming Anarchy," *Atlantic Monthly*, February 1994, 44–77.

20. Kaplan, "The Coming Anarchy," quoted in Philip Marfleet, *Refugees in a Global Era* (New York: Palgrave Macmillan, 2006), 3. See also Samuel Huntington, "The Clash of Civilizations?" *Foreign Affairs* 72: 3 (1993), 22–49.

21. Many recent films on immigration and refugees are outstanding, include the Dardenne Brothers' *La Promesse* (1996), Joshua Marston's *Maria Full of Grace* (2004), Nick Broomfield's *Ghosts* (2006), Thomas McCarthy's *The Visitor* (2008), Alfonso Cuarón's futuristic *Children of Men* (2006), and Neill Blomkamp's science fiction *District 9* (2009).

Objections to the argument that environmental change was causing profound population displacement quickly emerged from a variety of directions. For the sake of clarity, the objections can be divided into three main perspectives.

The first challenged the lack of analytic purchase. After the introduction of "environmental refugees" in the '80s, went this argument, the concept became accepted as a phenomenon although analysts had not examined systematically whether migration was caused by environmental degradation or by other factors. Conflict within a given region might prompt population movement that, in turn, stresses land capacity and resource management. Or government policy encouraging the growth of a certain kind of crop for export might prompt deforestation or desertification that, again, could disrupt traditional patterns. Critics argued that environmental factors might prompt population movements, but only as an intervening or contributing variable in more complicated causal chains.

Additional problems emerged with the tendency for circular citations, wherein one scholar might offer an analysis and approximation of expected migration flows, only to have the estimate accepted uncritically by subsequent scholars. In turn, fearing that their estimations might be too low, subsequent scholars increased their assessments, often based on little real evidence. In a form of the "precautionary principle"—taking preventive steps in case the worst-case scenario proves true—it was seen as better to err on the side of an inflated number than one too small. The result may have been an upward spiral of imprecise estimates.

The second objection to the use of CIM, especially in the context of refugee studies, stems from the lack of recognition in refugee jurisprudence. The heart and soul of refugee protection is the 1951 United Nations Convention Relating to the Status of Refugees, which defines a refugee as a person who:

> owing to a well-founded fear of being persecuted for reasons of race, religion, nationality, membership of a particular social group or political opinion, is outside the country of his nationality and is unable or, owing to such fear, is unwilling to avail himself of the protection of that country; or who, not having a nationality and being outside the country of his former habitual residence as a result of such events, is unable or, owing to such fear, is unwilling to return to it.[22]

22. The text is available at www.unhcr.org/protect/PROTECTION/3b66c2aa10.pdf.

The 1951 Refugee Convention emerged in the immediate aftermath World War II and reflected the preoccupations of the time, when m refugees were European in origin. Its assumptions were explic narrow, framing refugees as politically persecuted individuals seeking asylum. It was expanded in a 1967 protocol to offer protection to refugees outside the European context, which was ambitious, as governments were already seeking to restrict refugee protection. Certainly by the '00s, especially after September 11, securing protection for "traditional refugees" was challenging enough. Scholars and advocates for refugees argued that adding environmental refugees to the mix would be still harder.[23]

In a provocative 1985 essay, however, Shacknove asked, "Who is a refugee?"[24] Breaking with the narrow specification of a refugee in the Cold War context and grounding his answer to the question in the social contract theory of Hobbes, Locke, and Rousseau, Shacknove points out that refugees were traditionally understood as emerging from a context of persecution by the state or a "predatory sovereign." He argues for an expansion of the notion of "refugeehood" to include vital (economic) subsistence and protection from natural calamities, since they, too, compromise physical security. For Shacknove, then, "Refugees are, in essence, persons whose basic needs are unprotected by their country of origin, who have no remaining recourse other than to seek international restitution of their needs, and who are so situated that international assistance is possible."[25] Again, it is an expansive category. But the tripping mechanism is that, in an international system of sovereign states, if people must leave their country because basic rights are not being met, then they are seeking refuge. Staying in the country is unacceptable because of a "well-founded fear" that the government will not or cannot meet their needs. This tension between offering protection to people who are fleeing economic deprivation versus offering protection *only* to those who are moving because of political persecution remains profound. Shacknove's essay anticipated post–Cold War analyses that

23. Gil Loescher, *Beyond Charity: International Cooperation and the Global Refugee Crisis* (New York: Oxford University Press, 1993); Stephen Castles, *Environmental Change and Forced Migration: Making Sense of the Debate* (Geneva, Switzerland: UNHCR Evaluation and Policy Analysis Unit, 2002); Richard Black, "Fifty Years of Refugee Studies: From Theory to Policy," *International Migration Review* 35: 1 (Spring 2001), 57–78; and Richard Black, *Environmental Refugees: Myth Or Reality?* (Geneva, Switzerland: UNHCR Evaluation and Policy Analysis Unit, 2001).
24. Andrew Shacknove, "Who Is a Refugee?" *Ethics* 95: 2 (1985), 274–284.
25. Ibid., 277.

challenged the convention of characterizing refugees as simply "white, male anticommunists."[26]

The third objection to CIM was that it might have some ironic, even perverse and unintended, implications. In this line of argument, concerns about climate migrants and refugees seeking to cross international borders are counterproductive, playing into the hands of governments seeking to secure their borders and tighten refugee controls. In the spirit of pointing to a growing phenomenon, and even being inclined to help solve the challenges posed by forced displacement by environmental factors, analysts unwittingly give fuel to security-minded officials and electorates. If refugees motivated by environmental change are headed toward North Atlantic countries, so the reasoning might go, then efforts should be made to stop them. For Kibreab:

> The concept of the environmental refugee is increasingly used by states to justify restrictive refugee policies. Many academics, instead of questioning the conceptual and legal foundation of the term, have been using it uncritically, and so unintentionally contributing to the hardening of attitudes and policies against involuntary migrants.[27]

In a related dynamic, by attributing refugee flows to climate change, other causal factors might be left aside. Conflict, poor government policy, corruption, warfare, and nefarious international involvement in a region can prompt environmental degradation. Cold War involvement in the Horn of Africa during the '70s and '80s, or strategic efforts in Sahelian Africa as part of the Global War on Terror (GWOT) after September 11, may initiate displacement as much as, or more than, climate change. Pointing to environmental change alone lets policy makers off the hook. The invocation of climate refugees to justify a security response based on environmental *conflict* rather than on a broader, more people-centric notion of environmental *security* is perhaps even more prominent in recent

26. See B. S. Chimni, "The Geopolitics of Refugee Studies: A View from the South," *Journal of Refugee Studies* 11 (1998), 350–374; Aristide Zolberg, "Beyond the Crisis," in *Global Migrants, Global Refugees: Problems and Solutions*, eds. Zolberg and Benda (New York: Berghahn, 2001), 1–16; Stephen Castles, "Towards a Sociology of Forced Migration and Social Transformation," *Sociology* 37:13, 2003, 13–34; Susan Forbes Martin et al., *The Uprooted: Improving Humanitarian Responses to Forced Migration* (Lanham, MD: Rowman & Littlefield, 2005); and Caroline Moorehead, *Human Cargo: A Journey among Refugees* (New York: Henry Holt, 2005).

27. Kibreab, "Environmental Causes and Consequences of Migration: A Search for the Meaning of 'Environmental Refugee,'" 21.

years.[28] As examined at length in chapter 3, North Atlantic officials increasingly point to climate change as a security threat, with growth in CIM in coming decades as the catalyst for a policy response.

Finally, the politics of securitizing CIM raises issues concerning transit states—countries on the periphery of advanced-industrialized societies. Transit states occupy a pivotal position within broader migration systems. In their own interest, they seek to selectively and strategically encourage the emigration of their own citizens and to discourage immigration from elsewhere.[29] Meanwhile, North Atlantic destination countries urge these bordering states to assist them in controlling peoples seeking entry. For countries such as Mexico, Morocco, Tunisia, Libya, Albania, and Turkey, this role has transformed diplomacy and governance. This issue is examined directly in chapter 4.

CONCEPTUAL CHALLENGES AND ESTIMATES

Specifying the factors that might impel migration also remains a typological challenge.[30] An individual seeking to flee a cataclysmic event such as the volcanic eruption in Montserrat that began in 1995, or Central America's Hurricane Mitch in 1998, or the 2010 earthquake in Port-au-Prince, is in a different position than a second individual seeking to escape an ecosystem in decline because of willful policies such as dam building or strategic defoliation during war. It is a third migrant, one pressed to move because of unintentional but very real gradual climate degradation, that forms the heart of the concern here. While the degradation that motivates this third migrant is fundamentally anthropogenic, it is not the outcome of deliberate policy decisions. And its impact is often gradual, certainly not as acute or startling as that which drives the first two kinds of refugees. Table 1.1 seeks to clarify these distinctions, even as there may be empirical overlap.[31]

28. Nicole Detraz and Michele Betsill, "Climate Change and Environmental Security: For Whom the Discourse Shifts," *International Studies Perspectives* 10 (2009), 303–320.

29. For a treatment of the notion of "migration systems," see Castles and Miller, *The Age of Migration: International Population Movements in the Modern World*; and Pierette Hondagneu-Sotelo, *Gendered Transitions: Mexican Experiences of Immigration* (Berkeley: University of California Press, 1995).

30. Diane C. Bates, "Environmental Refugees? Classifying Human Migrations Caused by Environmental Change," *Population and Environment* 23: 5 (May 2002), 465–477.

31. Adapted and modified from Bates, "Environmental Refugees? Classifying Human Migrations Caused by Environmental Change."

Table 1.1 TYPOLOGY OF ENVIRONMENTAL REFUGEES

	I Disaster Unintended, catastrophic event		II Expropriation Willful, purposeful destruction		III Deterioration Incremental	
Sub-Category	Natural	Technological	Development	Ecocide	GHG	Depletion
General Example	Hurricane	Nuclear Disaster	Dam building	Defoliation	Climate change and island inundation	Deforestation
Empirical Example	Katrina 2005	TMI 1979	Three Gorges 2006	Agent Orange 1963–1969	Carteret Islands	Amazonia

As with any typology, table 1.1 presents conceptual challenges. For example, Type I events may, indeed, have climate connections. There is significant evidence that hurricanes will increase in frequency and intensity because of anthropogenic climate change.[32] And climate change may not be as gradual as the Type III category suggests. There would be little danger of mass migration flows if change were gradual, which would allow people to adapt. But climate models show the very real potential for punctuated and extreme spikes in climate change—that is, more dramatic events such as drought and flooding.[33]

To complicate the picture further, in all three types the individual's volition is questionable. It is often assumed that economic migrants have a degree of agency. They have made a calculated choice to move. In some instances, although it would be naive to overstate the case, there also seems to be an assumption that political refugees have made a strategic decision to flee or go into exile. The sense of volition is crucial to how and whether relief or *refugee* status should be accorded to the person in flight. In other words, the challenge for a Type III category—what is considered CIM in this analysis—is that more often than not, the forced migration is less obviously cataclysmic and shocking than the natural calamity or human-made disaster that motivates Type I refugees. For that matter, Type II refugees often have little opportunity to pause before they have to mobilize; relocation by a state as it builds a dam is obviously coerced and

32. Stanley Goldenberg et al., "The Recent Increase in Atlantic Hurricane Activity: Causes and Implications," *Science* 293 (2001), 474–479.

33. Timothy M. Lenton et al., "Tipping Elements in the Earth's Climate System," *Proceedings of the National Academy of Sciences* 105: 6 (February 12, 2008), 1786–1793; and National Research Council, *Abrupt Climate Change: Inevitable Surprises* (Washington, DC: National Academy Press, 2002).

Table 1.2 MATRIX OF CAUSE AND VOLITION

	VOLITION		
CAUSE	Voluntary	Compelled	Involuntary
Economic	Migrant		Trafficked individual
Environmental		Climate-induced migrant – Person displaced by climate change (Type III)	Person displaced by cataclysm or state policy (Types I or II)
Political			Traditional refugees (1951 Convention)

rarely gentle. These concerns become central, as Type III individuals, families, and communities contend with decisions on how to adapt to climate change. Table 1.2 presents these dynamics in a matrix, as a cross of volition versus causes. The empty blocks are provocative, begging debates about labels and their conceptual boundaries.

Migrants impelled to move because of graduate climate degradation sometimes act with intention. In fact, in studies associated with climate change, the mantra of the day is increasingly redolent of real estate's preoccupation with location: adaptation, adaptation, adaptation. Some of the adaptive strategies individuals employ are indeed quite willful and calculated and can include migration. Household calculations are made for some members of the family to move abroad, others to move to cities, and still others to stay behind.

So to argue that Type III refugees all suffer acute dislocation is to ignore empirical reality. Table 1.3 shows the continuum of volition.

Some CIM is voluntary, some fully forced. Most is somewhere in between. Even for coastal flooding scenarios, residents of low-lying areas might witness several seasons of inundations before seeking higher ground. The Carteret Islands, part of Papua New Guinea, have exhibited this dynamic, with some of the older members of the community preferring to remain behind as younger ones relocate.[34]

Finally, considering the range of predictions of future climate refugees is instructive. While there may be relatively precise estimates associated with Type I and Type II refugees and specific catastrophes, for Type III

34. Neil MacFarquhar, "Refugees Join List of Climate-Change Issues," *New York Times*, May 28, 2009; and Charles Hanley, "If an Island State Vanishes, Is It Still a Nation?" *Washington Post*, December 6, 2010.

Table 1.3 PERSON DISPLACED BY CLIMATE CHANGE (TYPE III)

Volition	Voluntary	Compelled	Forced
Category	Environmentally motivated migrant	Environmentally forced migrant	Environmental refugee

refugees predictions are wide and disparate.[35] There is also a conceptual muddle concerning the identification of "populations at risk" versus "populations displaced." The most cited figure is Myers's estimate of 200 million Type III refugees, although he leaves the time frame imprecise.[36] Such uncertainty about timing is actually refreshing given the challenges of predicting future scenarios. Some estimates are devoted to current accounting, others focus on 2030, and still others are geared toward 2050. The International Organization of Migration (IOM), like Myers, offers a relatively sober estimate of 200 million climate refugees but specifies the date of 2050.[37] The UK-based NGO Christian Aid gives what appears to be the largest estimate: 1 billion people displaced by climate change. The date offered is imprecise.[38]

Obvious problems persist in these estimates. In addition to the political, empirical, and definitional challenges noted thus far, there is the impossibility of comparing data sets. There are also assumptions about the rationality of decisions to migrate. As noted, individuals make decisions to migrate not always as strictly rational calculators, but as part of larger household, generational, and community structures.

Therefore, near-term estimates for specific migration systems are preferred here. As chapter 2 shows, paying attention to one system—Sahelian and sub-Saharan Africa—is daunting enough. Long-term estimates are fascinating to explore for their meta-level politics, though. Such issues form the heart of the recent volume *Sex, Drugs, and Body Counts*, which

35. Koko Warner et al., *In Search of Shelter: Mapping the Effects of Climate Change on Human Migration and Displacement* (New York: CARE International, 2009).

36. Norman Myers, "Environmental Refugees: A Growing Phenomenon of the 21st Century," *Philosophical Transactions of the Royal Society* 356 (2001).

37. International Organization for Migration, *Climate Change, Environmental Degradation and Migration: Addressing Vulnerabilities and Harnessing Opportunities: Discussion Note on Migration and the Environment MC/INF/288*, (Geneva, Switzerland: International Organization for Migration, 2008).

38. Christian Aid, *Human Tide: The Real Migration Crisis* (London: Christian Aid, 2007).

examines the ways in which the media and policy makers deliberately and/or naively accept politicized and questionable estimates.[39]

These concerns in mind, the working definition offered by the International Organization for Migration remains a common point of departure:

> Environmental migrants are persons or groups of person who, for compelling reasons of sudden or progressive changes in the environment that adversely affect their lives or living conditions, are obliged to leave their habitual homes, or choose to do so, either temporarily or permanently, and who move either within their country or abroad.[40]

As with any definition—Weber's oft-quoted definition of the state comes to mind—the constituent phrases are packed with meaning. Plus, its inclusiveness is a challenge: it encompasses Type I, Type II, and Type III refugees, in addition to IDPs. The UN High Commissioner for Refugees (UNHCR) has vigorously criticized the IOM definition as too broad to be useful, arguing that calling an IDP a migrant risks clouding crucial international legal terminology that reserves the label for someone living and working in another country.

CONCLUSION

There is ample conceptual muddle regarding CIM. This does not mean that superior analyses are not available; subsequent chapters explore such offerings at length. Nonetheless, the lack of definition and the incapacity to agree on fundamental methodologies, estimates, and scope provide the political opportunity for inappropriate and ineffective policy responses. The invocation of present and future refugee crises stemming from climate change can be politically loaded, and the potential for political misuse is profound. Pickering constructed a parody of newspaper articles using actual phrases from two respected Australian dailies, the *Sydney Morning Herald* and the *Brisbane Courier Mail*:

> "We" are soon to be "awash," "swamped," "weathering the influx," of "waves," "latest waves," "more waves," "tides," "floods," "migratory flood," "mass exodus" of "aliens," "queue jumpers," "illegal immigrants," "people smugglers," "boat

---

39. Peter Andreas and Kelly Greenhill, eds., *Sex, Drugs and Body Counts: The Politics of Numbers in Global Crime and Conflict* (Ithaca, NY: Cornell University Press, 2010).
40. IOM, *Climate Change, Environmental Degradation and Migration.*

people," "jumbo people," "jetloads of illegals," "illegal foreigners," "bogus" and "phoney" applicants and "hungry Asians" upon "our shores," "isolated coastlines" and "deserted beaches" that make up the "promised land," the "land of hope," the "lucky country," "heaven," "the good life," "dream destination," and they continue to "slip through," "sneak in," "gathering to the north," "invade" with "false papers," or "no papers," "exotic diseases," "sicknesses," as part of "gangs," "criminal gangs," "triads," "organized crime," and "Asian crime."[41]

Alas, there is little reason to suggest that similar language will not be used in the future. To emphasize climate change as contributing to over-whelming refugee flows potentially strengthens the hand of official and popular responses characterized by a narrow notion of security.[42]

International labor migration, often lamentably ignored in the main-stream globalization literature that emerged in the '90s, actually reveals the fundamental character of a global economy. Similarly, the existence of forced migration—of refugee crises, camps, and humanitarian relief efforts—is often cast as a failure of international society and a compromise of its integrity. For Agamben, individuals outside the boundaries of sover-eignty occupy a position of "bare life." In a state of "clandestinity," migrants have to "make themselves absent from the spaces they occupy."[43] These are important insights. Yet, arguably, migrants, refugees, and *clandestinos* actually reveal the full nature of the international system, and only by engaging them in theory and practice can we achieve a more complete understanding of international politics.[44]

The deeper challenges, therefore, are to (1) understand more fully the empirical realities that affect migration patterns, (2) think through how to "desecuritize" the discourse and policy associated with CIM, and (3) craft an ethical and practical set of policy initiatives to address climate change and the migratory flows to which it might contribute. The international system's response to the expected growth of transborder CIM is already a concern, and one that will only deepen in the years to come. Before the politics of CIM can be fully appreciated, however, a systematic appraisal of the science concerning climate change and migration in a specific context is essential.

41. Pickering quoted in Marfleet, *Refugees in a Global Era*, 312.

42. Betsy Hartmann, "Rethinking Climate Refugees and Climate Conflict: Rhetoric, Reality and the Politics of Policy Discourse," *Journal of International Development* 22 (2010), 233–246.

43. Susan Bibler Coultin, "Being En Route," *American Anthropologist* 107: 2 (2005), 195–206.

44. Emma Haddad, *The Refugee in International Society: Between Sovereigns* (New York: Cambridge University Press, 2008); and Nicholas Wheeler, *Saving Strangers: Humani-tarian Intervention in International Society* (New York: Oxford University Press, 2003).

CHAPTER 2

Scope and Dimensions

The Case of Sahelian and Sub-Saharan African
Migration to Europe

On February 12, 2002, then U.S. Secretary of Defense Donald Rumsfeld held a press conference on the nascent war in Afghanistan. The briefing is renown because of Rumsfeld's eerie comments on the presence of WMD in Iraq and the way in which he channeled Socrates' wise admission of ignorance:

> As we know, there are known knowns; there are things we know we know. We also know there are known unknowns; that is to say we know there are some things we do not know. But there are also unknown unknowns—the ones we don't know we don't know.[1]

Regardless of his failings as a secretary, Rumsfeld encapsulated the scientific project and the building of knowledge. Social scientists are especially adept at identifying the secretary's second category, "known unknowns," in the form of calling for further research and inquiry: "More investigation needs to be done in the areas of X, and questions persist about Y." It is a common way of closing a piece of research or a public presentation. It is even an opportunity to articulate a personal research agenda. Often the implication is, "I intend to turn to these matters in my own scholarship."

1. The transcript is available at www.defense.gov/transcripts/transcript.aspx? transcriptid=2636.

This habit has been especially pronounced with regards to climate-induced migration because it has been so contested as a concept. Since the mid-'80s, the ritualistic "more work needs to be done" incantation has often appeared in analyses of CIM. One might argue that it serves different rhetorical purposes, depending on the political position and agenda of the author. For the security alarmists examined in chapter 3, it is a way to signal that there is a real problem that merits real concern. For those motivated by ethics and concern for the humanitarian plight of migrants pushed to move by climate change, or calling for a global governance perspective, as discussed in chapter 5, it is a call to action. For methodologists keen on understanding the role that climate change plays in demographic pressures, it is born of a demand for rigor, testability, and enhancement of our knowledge.

The problem, again, is that calculations about the scope of CIM have often been back-of-the-envelope, often space- and time-bound, and based on other estimates. Predictions and scenario building, however sophisticated, are also fraught with complications, as analysts are caught between product-oriented and process-oriented exercises.[2] In the product-oriented dimension, the content of the scenario—the outcome anticipated and the interrelationships between causal factors—remains foremost. The main question seems to be how the expected change can be dealt with. As a result, the expected outcome of the scenario takes on a life of its own: the future is written. The world is going to get warmer, and it will drive many people to migrate. It is the way it is going to be: deal with it.

In a process-oriented perspective, by contrast, the activity of scenario development is the goal. A scenario is "a means to motivate . . . learning, find commonalities across different perspectives, achieve consensus on goals, or come to a shared understanding of challenges."[3] Process-oriented scenario building creates a product, of course. But the emphasis is on the constitutive nature of the interaction in which the scenario was formed. Participants creating ideas about the future also can brainstorm about how to avoid worst-case scenarios. In this regard, the goal is to speculate about possible future scenarios as a learning process and, perhaps, a means of averting worst cases. Indulging in tragic thinking can be a way of avoiding tragedy.

2. Simone Pulver and Stacy VanDeveer, "'Thinking about Tomorrows': Scenarios, Global Environmental Politics, and Social Science Scholarship," *Global Environmental Politics* 9: 2 (May 2009), 1–13.

3. Brian O'Neill et al., "Where Next with Global Environmental Scenarios?" *Environmental Research Letters* 3: 4 (2008), 1.

Bearing in mind this distinction, the success of scenario bu becomes an issue for evaluation. What is a given scenario's success: dictive accuracy? Its ability to generate good decisions? The ability (ticipants to learn from the process? The first two question product-oriented; the third is process-oriented. The purpose of scenario building—especially its process-oriented dimension—must be kept in mind, or the fact that it simultaneously shapes and is embedded in social contexts is forgotten. Assumptions about the way things are and the way things will be in the future quickly become reinforced, as do assumptions about the improbability of change.

"Thinking about tomorrow," therefore, prompts us to think about immediate and short-term policy efforts that can make the future better—that will prevent the realization of a worse-case scenario. Put bluntly, the worst-case scenario is acceptance of the worst-case scenario as a given.

All this is obviously relevant in looking at CIM, its past dynamics, and its anticipated future scope. So much of the current discourse associated with CIM and climate change is dire, lending an air of inevitability. It is more useful, instead, to examine the range of future scenarios as a means of prompting a critical evaluation of policy alternatives. As examined in chapters 3 and 4, the emergent security discourse and support for the authoritarian attributes of transit states is shortsighted. Preferable, as examined in chapter 5, is the "desecuritization" of discourse: a more process-oriented and cooperative emphasis on GHG mitigation, adaptation, and protection of vulnerable populations. Governance at a wide array of levels is vital.

This chapter, after an analysis of anthropogenic climate change, offers a systematic appraisal of the state of scientific knowledge vis-à-vis CIM. It remains cognizant of the evidence from around the world—the Pacific islands, South Asia, the Caribbean—yet grounds the analysis in Africa. Specifically, it explores the findings concerning the impact of climate change on migration patterns in the Sahel and sub-Sahara. This is ostensibly the primary region of origin for the mixed flow of migrants to the Maghreb, the ultimate destination being the European mainland. And it is expected to remain the primary sending region for Europe. Climate change has contributed, in complicated and multifaceted ways, to demographic pressures, and this will accelerate in the decades to come. At the same time, it is essential to avoid panic. CIM will likely intensify, yes, but perhaps not as alarmingly as some suggest. Flows to the Maghreb and Europe will increase, too, but much of the exacerbated CIM will likely remain regional, rural–urban, and south–south. The bulk of migrations will remain south of the Sahara and within the African continent. A Malian villager outside the city of Mopti may endeavor to travel to a transit state such as Tunisia, with

Italy as a final destination, but she is more apt to travel to Bamako as part of a household or communal strategy. She might try to travel to another Sahelian city, such as Niamey or N'Djamena, but a coastal city such as Abidjan, Accra, Cotonou, or Lagos is a more likely destination. This picture is quite different from that painted by North Atlantic analysts, policy makers, and the media. The chapter concludes with an assessment of potential scenarios concerning migration flows toward the Maghreb and Europe in the next two decades.

ANTHROPOGENIC CLIMATE CHANGE

Before turning to the impact of climate change on population movements, it is pertinent to consider climate change itself and to ask directly why the Earth is warming, to what extent climate change is anthropogenic, and whether it is expected to continue.

The Earth is warmed by the absorption of radiation from the Sun and cooled by the emission of infrared radiation back into space. Most of the Earth's infrared radiation is absorbed and reemitted by gases, especially in the troposphere, which extends from the surface to 10 to 15 km (higher in the tropics and lower at the poles).[4] The result is a "greenhouse effect" that warms the surface. Without this effect, first theorized by Joseph Fourier in 1820, Earth would be too cold to support life. The principal greenhouse gases are H_2O and CO_2, although other gases such as CH_4, chlorofluorocarbons and N_2O are crucial, too. An additional complication is the role of clouds and aerosols. For example, clouds are implicated in both cooling and warming; low clouds tend to cool the surface, while high, thin clouds contribute to warming. Human-made and natural aerosols (such as sand and airborne sea salt) also have complicated dynamics; some trap heat against the Earth, while others contribute to the albedo effect—the reflection of solar radiation back into space. Volcanic eruptions contribute to climate change, too, by releasing gases and aerosols that often cause cooling.

A big problem is that a greenhouse is actually a rather ridiculous metaphor. The word *greenhouse* conjures an image of a bright, bucolic room with a wonderful smell, carefully arranged tables of lovely flowers and plants, coils of watering hose, and a solid concrete floor with drainage. The Earth is nothing of the sort. It is big and round. It possesses vast, salty oceans that redistribute solar energy from the equator to the poles. The salinity and differential temperature of the oceans are central factors because of their

4. L. R. Kump et al., *The Earth System* (Upper Saddle River, NJ: Prentice-Hall, 2004).

impact on density and thermohaline circulations, which move parcels of water up and down the water column.[5] Moreover, the Earth spins on an axis. Every day. Quickly. This spinning induces the Coriolis force, which affects everything: ocean currents, geostrophic thermal winds (jet streams, for example), atmospheric circulation systems such as the Hadley cells and the Walker cells implicated in the El Niño-Southern Oscillation (ENSO), and even the navigational calculations of airline pilots.[6]

Additionally, the Earth orbits the Sun in an eccentric ellipsis, with its axis tilted in relation to the plane of the orbit. Piling on even more complexity are the findings of Johannes Kepler in the seventeenth century, deepened by Milutin Milankovitch in the 1920s, that the elliptical nature of the orbit, the tilt of the axis, and the precession (wobble) of the axis display a periodicity that can be tens of thousands and even hundreds of thousands of years in length.[7] As an example, the North Pole is currently pointed at Polaris, or the North Star, and remains so even as the Earth orbits the Sun. However, the spin axis wobbles. So 5,000 years ago the Egyptian pyramids were oriented toward a different north star, Alpha Draconis, and 13,000 years ago Vega was the north star. Similarly, the elliptical nature of the Earth's orbit around the Sun is affected by the gravitational pull of other planets; it was different thousands of years ago.

There is also the fact that the amount of solar energy, or insolation, varies at different latitudes. There is more at the equator and less toward the poles. And insolation, in turn, depends on the character of the Earth's precession, obliquity (tilt), and eccentricity (orbit). The eccentricity today is low, meaning that there is little opportunity for long winters that facilitate glaciation, and it is expected to stay low for 30,000 more years. (The Earth came out of the last ice age, the Pleistocene era, 10,000 years ago; it is now in the Holocene era.) Thus, the unusually cold winters in the Northern Hemisphere needed to form glaciers are not occurring. Even if there were not an increase in GHGs due to human activity, the interglacial period would persist.[8]

And that is the additional problem. Because of human activity since the eighteenth century, the buildup of GHGs in the atmosphere is increasing rapidly, especially beginning in the twentieth century. This is accelerating, okay, let's call it the greenhouse effect. Some have even argued that

5. Stefan Rahmstorf, "Ocean Circulation and Climate during the Past 120,000 Years," *Nature* 419 (September 12, 2002), 207–214.

6. A seminal work on ENSO, building on Jacob Bjerknes's theories, is Mark Cane et al., "Experimental Forecasts of El Niño," *Nature* 321 (June 26, 1986).

7. Kump, *The Earth System*, 276.

8. Ibid.

the Holocene era has been supplanted by the "Anthropocene era."[9] In other words, humans are now affecting the Earth to such an extent that, in the context of geologic "deep time," we define the era. Especially striking is that the scientific capacity to begin to understand this is very recent. To return to the aforementioned ENSO, Peruvian fishing communities have known for centuries that naturally occurring deep-sea upwelling, which supplies nutrient-rich cold H_2O to fisheries, can periodically diminish at Christmastime—hence the term El Niño, or "the Christ Child." Such local knowledge is profoundly useful, as discussed below with respect to Africa. Still, it is only since the '80s, with the advent of satellite-imaging technology and a full array of ocean-monitoring buoys, that the full complication of the seesaw of atmospheric pressure and moisture— and its relationship to sea level pressure (SLP) and sea surface temperature (SST)—has become better understood. And the impact of elevated GHGs on future ENSO dynamics is even more problematic and the subject of intense analysis.[10]

The abiding point is that the evidence is clear that human-induced GHGs are contributing to and exacerbating naturally occurring climate change.[11] There is natural variability that operates on decadal scales. And within decadal changes there is often significant fluctuation, or "noise," that makes it challenging to clearly discern the direction, or "signal," of climate change. Nonetheless, warming continues. Oreskes reviewed 928 scientific paper abstracts published in peer-reviewed journals between 1993 and 2003 and concluded that the scientific consensus is clear.[12] There are difference of opinions on the amount, the direction, the regional differences, and future projections, but the overall view is well understood. Because the 2007 Fourth Assessment Report (AR4) of the IPCC was compiled and redacted in 2006, its findings have already become somewhat outmoded. A 2009 updating, building on Working Group I or the

9. See Dipesh Chakrabarty, "The Climate of History," *Critical Inquiry* 35 (Winter 2009), 197–222.

10. Gabriel A. Vecchi and Brian J. Soden, "Global Warming and the Weakening of the Tropical Circulation," *Journal of Climate* 20 (September 1, 2007), 4316–4340; Mark Cane, "The Evolution of El Niño, Past and Future," *Earth and Planetary Science Letters* 230 (2005), 227–240.

11. Ian Allison et al., *The Copenhagen Diagnosis: Updating the World on the Latest Climate Science* (Sydney, Australia: University of New South Wales Climate Change Research Center [CCRC], 2009); and James Hansen, "Defusing the Global Warming Time Bomb," *Scientific American*, March 2004, 68–77.

12. Naomi Oreskes, "The Scientific Consensus on Climate Change: How Do We Know We're Not Wrong?" in *Climate Change: What It Means for Us, Our Children, and Our Grandchildren*, eds. Joesph F. C. Dimento and Pamela Doughman (New York: Cambridge University Press, 2007), 73.

"Physical Science Basis," was prepared for the COP15 in Copenhagen.[1'] found the following:

- CO_2 emissions in 2008 were nearly 40 percent higher than in 1990. Even if global emission rates were stabilized at present-day levels, 20 more years of emissions would cause warming to exceed 2°C in 2030.
- Despite fluctuations, the trend continues to be one of warming over the last 25 years.
- Ice sheets, glaciers, and ice caps are melting at accelerating rates, as is Arctic sea ice.
- Sea level is rising faster than estimated—about 80 percent above the IPCC prediction. By 2100, global sea level is likely to rise at least twice as much as predicted by the AR4.

Climate change has happened, it is happening, and it will continue to happen. And humans are contributing to it in complicated ways.

Finally, as noted, abundant effort is made to project climate change into the future. This comes up against tensions between projections, prediction, and forecasting. One often hears the sardonic question, how can climate change be predicted when forecasters cannot say with certainty if it'll snow next week? This question reflects confusion over climate modelers' efforts to project decadal variability, and how that task is vastly different from predicting next week's weather. Using complex coupled atmospheric and oceanic general circulation models (AOGCMs)—based on the conservation of mass, energy, and momentum in the atmosphere and ocean, and the physical processes involved in the interrelationships between them—fairly sophisticated projections can be achieved. If, then, the average of several models is taken, one can detect a clear signal. The evolution of the AOGCM models is important to bear in mind, too. In the early '70s, they were rather rudimentary and focused only on the atmosphere. By the early '90s, however, the models had integrated land-surface, ocean, and sea-ice components with the atmospheric parameters. Finally, by the early '00s, AOGCMs included atmosphere, land surface, ocean ice; sulphate aerosols; the carbon cycle; dynamic vegetative cover; and atmospheric chemistry. Refinement continues. The results are very credible, as past projections are consistently tested against actual events. Climate scientists may not be able to say what the *weather* will be on March 22, 2030, but they can project with increasing accuracy what the *climate* will be.

13. Allison et al., *The Copenhagen Diagnosis: Updating the World on the Latest Climate Science.*

Such efforts have also led to increasingly sophisticated analyses of tipping points, where abrupt, rather than gradual, climate change could occur.[14] A "tipping element" is a large-scale subcontinental system: Arctic summer sea ice, the West African monsoon (WAM) system, the Greenland ice sheet, the ENSO, or the Atlantic thermohaline circulation. Abrupt change in any one such system can prompt, in turn, profound changes in the Earth's climate. If because of unforeseen dynamics the Greenland ice-sheet melt accelerates faster than predicted, it could throw the climate out of balance much more rapidly. As Lenton et al. argue, "Society may be lulled into a false sense of security by smooth projections of global climate change."[15] What may happen is a rapid transition after a tipping point is passed.

Such issues also raise crucial questions associated with mitigation. Namely, how might it be possible to steer clear of critical tipping points? And with respect to adaptation, can the qualitative change after the critical point is passed be tolerated and/or adapted to?

PAST AND FUTURE CLIMATE CHANGE IN AFRICA

How climate change plays out (and is expected to manifest further) is different from region to region because of the complicated character of the Earth's oceans, landmasses, and atmosphere.[16] Some regions have been more profoundly affected by climate change, and different regions will be affected differently. But some generalities are offered with considerable confidence. The Northeast United States is likely to experience relatively less robust temperature increases yet greater precipitation. More frequent winter storms may arise, too, as Hadley cells expand northward because of a reduction in the equator-to-pole temperature gradient. The tropics will experience more frequent rainfall, with drying expanding at 30°N and 30°S, where the descending arm of the Hadley cells are located. Thus, the American Southwest, the southern rim of the Mediterranean, and South American, Australian, and Southern African deserts are likely to experience a sharp increase in temperature and a decline in precipitation.[17]

14. National Research Council, *Abrupt Climate Change: Inevitable Surprises* (Washington, DC: National Academy Press, 2002).

15. Timothy M. Lenton et al., "Tipping Elements in the Earth's Climate System," *Proceedings of the National Academy of Sciences* 105: 6 (February 12, 2008), 1792.

16. Gavin A. Schmidt, "The Physics of Climate Modeling," *Physics Today* (January 2007), 72–73.

17. Richard Seager et al., "Model Projections of an Imminent Transition to a More Arid Climate in Southwestern North America," *Science* 316 (May 25, 2007), 11081–11086.

Regional specifics remain open to question, though, because of the challenges of projections. Africa is especially challenging, in part because its size inhibits generalities. Mercator map projections have led to a common misperception that Greenland is the same size as the African continent. Mollweide or Behrmann projections provide a clearer sense of Africa's scale. Africa's 30.3 million square kilometers can hold China (9.6 million km^2), the United States (9.4 million km^2), Western Europe (4.9 million km^2), India (3.2 million km^2), and Argentina (2.8 million km^2) combined. The Democratic Republic of the Congo alone is larger than Western Europe.

Additionally, Africa is challenged by a lack of monitoring stations and, in turn, poor access to the data that is gathered. Monitoring stations on the order of what has existed for hundreds of years in the North Atlantic were not set up by European colonial authorities until the late 1920s, and then they were spotty. Many maps of global projections show the absence of data points for key portions of Africa. This has to change. Suffice to say, for now, obtaining data presents an ongoing challenge.

How has climate change affected Africa in recent decades? Funk et al. argue that climate models demonstrate that declines in eastern and southern African growing-season rainfall are linked to anthropogenic warming in the Indian Ocean.[18] The connections between ENSO events are not well elaborated in their analysis, although they do establish a connection between Indian Ocean warming and Pacific SST dynamics. Similarly, Trenberth et al., of Working Group I of the IPCC's AR4, argue that warming in recent decades has contributed to an earlier onset of the rainy season over northeastern Africa and a late start in southern Africa.[19] West Africa's multidecadal variability in rainfall has also been pronounced. For the Sahel, in particular, the rainfall and drought patterns during the '70s, '80s, and '90s are "among the largest climate changes anywhere." Despite some recovery from the worst of the '70s, the mean of the '90s is still below the pre-1970 level.[20] Biasutti and Giannini demonstrate that the Sahel's severe drying between the '50s and the '80s, with a partial recovery or "greening" since, is attributable to changing SST rather than anthropogenic land-surface modification such as land-use practices. They further argue that 30

18. Chris Funk et al., "Warming of the Indian Ocean Threatens Eastern and Southern African Food Security but Could Be Mitigated by Agricultural Development," *Proceedings of the National Academy of Sciences* 105: 32 (August 12, 2008), 11081–11086.

19. Kevin E. Trenberth et al., "Observations: Surface and Atmospheric Climate Change," in *Climate Change 2007: The Physical Science Basis. Contribution of Working Group I to the Fourth Assessment Report of the IPCC*, eds. S. Solomon et al. (New York: Cambridge University Press, 2007), 235–336.

20. Ibid.

·cent of the drying the Sahel experienced in the late twentieth century was attributable to external forcings such as GHGs and reflective aerosols.[21]

Such research facilitated a revision of the prevailing assumption that human activity had prompted the prolonged drought of the '70s and '80s in the Sahel. It is true that there is evidence that a human-induced reduction in vegetative cover could cause an increase in surface albedo, which in turn could reduce precipitation and lead to a further decrease in vegetative cover and enhancement of the albedo. Yet Giannini et al. demonstrate the role of oceans—both local (Atlantic and Indian) and remote (Pacific)—in forcing deep convection in the ITCZ to migrate seaward and away from the Sahel, thereby creating drought.[22]

As for the future, warming will affect African geological and biological systems. Rainfall patterns are likely to change, but it is crucial, again, to stress that the impact will be differential for the continent. Taking Africa in toto is highly problematic. As de Wit and Stankiewicz demonstrate in their superb analysis, projecting "a uniform change on a continental scale is a gross simplification."[23] Toulmin argues similarly.[24] For example, the continent's river drainage systems are useful proxies for demonstrating the complicated relations between precipitation and geological and biological systems. De Wit and Stankiewicz's analysis highlights the fact that Africa possesses 12 distinct river systems: the Nile, Senegal, Niger, Volta, Congo, Rufiji, Ganane, Zambezi, Okavango, Limpopo, and Orange river systems, and Lake Chad. Several of them are enormous and intricate. The Nile River, for example, starts in Uganda at the equator before emptying into the Mediterranean Sea at 31°N.

Further, de Wit and Stankiewicz divide Africa into three hydrological zones: less than 400 mm per year, between 400 mm and 1,000 mm, and greater than 1,000 mm. The dry areas already receive low rainfall and experience low perennial drainage. High-rainfall areas (greater than 1,000 mm per year) located at the equator's rain forests are projected to experience a 17 percent decrease in perennial drainage. This is significant and likely to cause stress, yet perhaps much less than one might assume. In some projections, by contrast, the equatorial regions are expected to

21. Michela Biasutti and Alessandra Giannini, "Robust Sahel Drying in Response to Late 20th Century Forcings," *Geophysical Research Letters* 33: L11706 (2006).

22. Alessandra Giannini et al., "Oceanic Forcing of Sahel Rainfall on Interannual to Interdecadal Time Scales," *Science* 302 (November 7, 2003), 1027–1030.

23. Maarten de Wit and Jacek Stankiewicz, "Changes in Water Supply across Africa with Predicted Climate Change," *Science* 311 (2006), 1917–1921.

24. Camilla Toulmin, *Climate Change in Africa* (London: Zed Books, 2010).

be wetter in a warming world. Warmer temperatures bring more precipitation because of moisture loading. According to the Clausius-Clapeyron relationship, as wetter parcels of air rise, they release the moisture as precipitation.[25]

It is the middle zone that especially concerns de Wit and Stankiewicz, as it comprises 25 percent of the continent and touches on 75 percent of its countries. The expected change in precipitation by the end of the twenty-first century (based on a composite of 21 AOGCMs) points up the need for special concern about southern Africa as well as Madagascar, and about the drying associated with the subtropics north of the Tropic of Cancer (23°N) and south of the Tropic of Capricorn (23°S). This includes the northern tier in the Mediterranean Basin. It is these regions that are expected to suffer the brunt of the drying. Sahelian and sub-Saharan Africa above the southern tier, on the other hand, are not predicted to be as brutally affected by climate change as conventional assertions suggest.[26]

As always, precise scientific determinations about the future are hard to make. For example, Funk et al.'s aforementioned conclusions about eastern Africa's recent decline in rainfall are at odds with IPCC assessments that anticipate precipitation increases. Christensen et al., of Working Group I of the AR4, argue, "There is likely to be an *increase* in annual mean rainfall in East Africa."[27] Nonetheless, Battisti and Naylor conclude that the average growing season temperatures by the end of the twenty-first century—and even earlier for some parts of the continent—can be expected to exceed the hottest recorded during the twentieth century.[28] They express high certainty that this will mean more frequent agricultural droughts—elevated evapotranspiration, low soil moisture, and high rates of H_2O runoff from hard soil when it does rain. In the end, the best prediction that can be achieved appears to be one that combines the tenor of de Wit and Stankiewicz's analysis—namely their use of AOGCM models and field station monitors—and local knowledge.

25. Kevin E. Trenberth et al., "The Changing Character of Precipitation," *Bulletin of the American Meteorological Society* 84 (September 2003), 1205–1217.

26. de Wit and Stankiewicz, "Changes in Water Supply across Africa with Predicted Climate Change."

27. Jens Hesselbjerg Christensen et al., "Regional Climate Projections," in *Climate Change 2007: The Physical Science Basis. Contribution of Working Group I to the Fourth Assessment Report of the Intergovernmental Panel on Climate Change*, eds. S. Solomon et al. (New York: Cambridge University Press, 2007), 850, emphasis added.

28. David S. Battisti and Rosamond L. Naylor, "Historical Warnings of Future Food Insecurity Unprecedented Seasonal Heat," *Science* 323 (January 9, 2009), 240–244.

A useful dichotomy for understanding the effect of climate change on Africa's population is that of impacts-led versus vulnerability-led approaches. Impacts-led approaches focus on the observed and expected impact of climate change on the continent and its subregions. The challenge is that the continent is vast and the specific effects of climate change remain the subject of intense scrutiny. And researchers often disagree. Sometimes a disparity in an impact or prospective impact can be explained by different methodological approaches, or in how data is initialized and bounded. Sometimes insights can be born of multimodel averages using macro-level data, with other studies using data obtained from local stakeholders.

Additionally, to discern change is one thing, but to attribute it to a specific cause is another. For example, separating anthropogenic climate change from human practices is challenging when discussing vegetative cover. The change in vegetation cover observed by satellite imaging is attributable to both climate change and changing land-use patterns or urbanization. In the '70s, the Charney hypothesis (after the work of Jule Charney) suggested that albedo increases because of human factors; overgrazing and cultivation lead to a general cooling of the land surface, subsequent reduction in evapotranspiration and convergence, and reductions in rainfall because of lack of clouds. This was challenged by further studies on the albedo effect and on the influence of remote oceanic SST.[29] Additionally, reduction in forested land might be caused by desertification, soil erosion, and salinization, but it might also be caused by population stressors or efforts to "marketize" firewood. Nonetheless, impact-led approaches focus on Africa's continental position, its scale, and its complexity in an attempt to specify the impact of climate change on the continent. Rosenzweig et al. of Working Group II of the AR4 argue that impacts-led approaches emphasize the detection and attribution of climate change with attention to changes in temperature and rainfall.[30]

Vulnerability-led approaches, by contrast, converge on Africa's circumstance in the international arena, its legacy of colonial rule, its tragic experience with Cold War and post–Cold War geopolitics, and its challenges with

29. Alessandra Giannini et al., "A Global Perspective on African Climate," *Climatic Change* 90 (2008), 359–383.

30. Cynthia Rosenzweig et al., "Assessment of Observed Changes and Responses in Natural and Managed Systems," in *Climate Change 2007: Impacts, Adaptation and Vulnerability: Contribution of Working Group II to the Fourth Assessment Report of the Intergovernmental Panel on Climate Change*, eds. M. L. Parry et al. (Cambridge, UK: Cambridge University Press, 2007), 79–131.

governance. The continent grapples with endemic poverty, limited access to capital, and significant population pressures.[31] (Emphasizing vulnerability brings to mind the distinction in the international political economy literature between a country's *sensitivity* to trade or currency fluctuations and the challenges of maintaining macroeconomic stability in order to avoid *vulnerability*.[32]) In other words, this line of reasoning goes, Africa's vulnerability is particularly acute, so it is more at risk from climate change that, theoretically, might be less devastating in other contexts. To take an example, climate change will aggravate the already acute water stress that parts of Africa experience. Freshwater runoff into river systems and reservoir basins is threatened not only by climate change but also by low levels of development and fast growing populations.[33] And meeting these challenges is going to be costly economically—whether by the implementation of adaptation strategies (such as the upgrade of infrastructure or the introduction of innovative technologies) or inaction (deterioration of water quality and crop yields). In tropical areas where water is relatively abundant, the challenge will be quality and provision. In arid and semiarid regions, the challenge will be scarcity.

Water insecurity throughout the continent makes Africa vulnerable to further climate change, but it is difficult to establish causality. Is Africa's climate harsher and, therefore, inclined to prompt acute stresses in vulnerable societies? Or do vulnerable societies exacerbate the problems caused by climate change, problems that might otherwise be surmounted? Examining the continent in a disaggregated fashion is crucial, too, as the impact will vary by region.

This raises still further questions. For example, is Africa truly as vulnerable as is often suggested? Mortimore, for example, takes exception to the prevailing wisdom that Africa is especially vulnerable or that, in the words of the Stern Review, "the continent's low adaptive capacity serves as a major constraint to her [sic] ability to adapt."[34] He argues that people in the Sahel are typically cast as poor and incompetent, lacking sufficient

---

31. Michel Boko et al., "Africa," in *Climate Change 2007: Impacts, Adaptation and Vulnerability: Contribution of Working Group II to the Fourth Assessment Report of the Intergovernmental Panel on Climate Change*, eds. M. L. Parry et al. (Cambridge, UK: Cambridge University Press, 2007), 433–467.

32. Robert O. Keohane and Joseph S. Nye, *Power and Interdependence*, 2nd ed. (Boston, MA: Little Brown, 1989).

33. Charles J. Vörösmarty et al., "Global Water Resources: Vulnerability from Climate Change and Population Growth," *Science* 289 (July 14, 2000), 284–288; and Hallie Eakin and Amy Lynd Luers, "Assessing the Vulnerability of Social-Environmental Systems," *Annual Review of Environmental Resources* 31 (2006), 365–394.

34. Quoted in Michael Mortimore, "Adapting to Drought in the Sahel: Lessons for Climate Change," *WIRES Climate Change* 1 (2010), 134–143. Referring in English to countries or to an entire continent by the feminine form of a pronoun, as the Stern Review does, is jarring.

local knowledge, and too rooted in their place. This is all contrary to the fact that people have adapted for millennia, are mobile, and exhibit crucial innovation. Mortimore further argues that adaptation to climate change cannot be separated from development within locally defined efforts. He avers:

> The continuing separation of climate change adaptation from development may be inefficient or damaging. A global priority for rushing adaptation to the top of the agenda may encourage the rebranding (and duplication) of conventional development knowledge and the neglect of development studies that are immediately relevant to adaptation.[35]

The importance of initiatives that emphasize development *and* environmental sustainability is addressed further in chapter 5.

Moreover, stipulating that Africa is especially vulnerable may undermine efforts to enhance the capacity to adapt. The continent faces deep challenges, but important research has been done on ways to improve data gathering, forecasting, and the dissemination of weather and climate information across the continent. As noted earlier, the continent starts from a low baseline of scientific knowledge. Washington et al. demonstrate that the coverage of weather stations in Africa remains quite spare.[36] Such structural disadvantages, in addition to the relatively poor training of African scientists, need to be remedied. There is also a gap in the integration of climate information into actual scientific knowledge.[37]

Positive prospects are evident as well. For example, in East Africa, pastoralists are quite adept at evaluating information presented to them. Such information often corroborates local wisdom—garnered from examining the intestines of slaughtered animals, bird flight patterns, clouds, livestock behavior, and so on—used to predict seasonal weather patterns.[38] (As noted earlier, for centuries Andean farmers have adjusted their potato planting according to the relative haziness of the Pleiades at the southern winter solstice. Researchers have ascertained that high cirrus clouds associated

35. Ibid.

36. Richard Washington et al., "African Climate Change: Taking the Shorter Route," *Bulletin of the American Meteorological Society* 87: 10 (2006), 1355–1366.

37. International Research Institute for Climate and Society et al., *A Gap Analysis for the Implementation of the Global Climate Observing System Programme in Africa, IRI-TR/06/1* (Palisades, NY: IRI, 2006).

38. Winnie Luseno et al., "Assessing the Value of Climate Forecast Information for Pastoralists: Evidence from Southern Ethiopia and Northern Kenya," *World Development* 31: 9 (2003), 1477–1494.

with El Niño create the haziness.[39]) This is time-honored knowledge, but Luseno et al. discern that pastoralists are willing to incorporate forecasting from external sources, albeit with limitations. Among those limitations is the forecasting uncertainty itself, as well as the spatiotemporal variability under which the herders must operate—what is true in one part of a plain may not be true elsewhere. In contrast to this sober yet relatively optimistic picture, Tarhule and Lamb conclude that farmers in West Africa do *not* integrate climate forecast data into their information as much as they could or should. In either case there is considerable need for improvement.[40]

Clearly—and consistent with the fundamental argument of this book that a security-minded approach to CIM diverts intellectual energy from more important endeavors—more effort must be devoted to both development *and* adaptation. For years, *adaptation* was a bad word in climate change circles because it was seen (1) as an admission by North Atlantic countries of liability and the need to offer assistance, and/or (2) as undermining efforts to pursue mitigation and abatement of GHG emissions.[41] Yet adaptation *must* be pursued in the context of development projects. Such projects should devote their efforts to climate forecasting and to a deepening of societal capacity to use the resulting information, such as by improving educational systems and the dissemination of information.

The ritualistic incantation that Africa is especially vulnerable may undermine the political will needed to improve the response by the continent and the wider international community. Moreover, given that African societies contribute only 2.5 percent of the world's CO_2 emissions, mitigation efforts within the continent are not especially pertinent. In his statement on behalf of African countries at Copenhagen's COP15 in December 2009, Ethiopian Prime Minister Meles Zenawi wrote:

> Africa lose[s] more not only because our ecology is more fragile but also because our best days are ahead and lack of agreement here could murder our future even before it is borne [*sic*]. Because we have more to lose than others, we have to be prepared to be flexible and be prepared to go the extra-mile [*sic*] to accommodate others.[42]

39. Benjamin S. Orlove et al., "Forecasting Andean Rainfall and Crop Yield from the Influence of El Niño on Pleiades Visibility," *Nature* 403: 6 (2000), 68–71.

40. Aondover Tarhule and Peter J. Lamb, "Climate Research and Seasonal Forecasting for West Africans," *Bulletin of the American Meteorological Society* 84 (2003), 1741–1759.

41. E. Lisa F. Schipper, "Conceptual History of Adaptation in the UNFCCC Process," *Review of European Community and International Environmental Law* 15: 1 (2006), 82–92.

42. Meles Zenawi, Statement on Behalf of the African Group at the COP 15 (Copenhagen, Denmark: 2009).

Obviously, Meles accepts the assumption that Africa is more fragile than other regions of the world, but his emphasis on innovative development strategies is crucial.[43]

As a brief but relevant aside before turning to migration issues, in a provocative essay about the blockbuster film *Avatar*, Klare proffers a screenplay treatment for filmmaker James Cameron's inevitable follow-up. In characteristically erudite fashion, Klare argues that Cameron should eschew a sequel and, instead, make a prequel. Klare suggests that it be about the Earth *before* wounded veteran Jake Scully leaves for Pandora. Klare envisions our planet in 2144 as characterized by tripolar competition: a North American Federation, Greater China, and the North European Alliance. The three resource-hungry blocs compete for resources from the rest of the world. This is an intriguing and provocative piece of "political science fiction," and a good read. Yet Klare offers a striking passage pertinent to the discussion here:

> Thanks to global warming, the planet's tropical and subtropical regions, including large parts of Africa, the Mediterranean basin, the Middle East, and South and Southeast Asia, as well as Mexico and the American Southwest, have become virtually uninhabitable. Many island nations and coastal areas, including much of Florida, Bangladesh, Vietnam, Sri Lanka, Indonesia, and the Philippines, lie under water.[44]

It's reasonable to conclude that Africa's climate will be different in 120 years, likely harsher for significant portions of its population. But to characterize Africa as "virtually uninhabitable"—even looking out 120 years—constitutes a profound (and increasingly commonplace) assumption. In the next century, hundreds of millions of people will of course be living in Africa, and they will adapt to climate change. They will *have* to. This is not to suggest that Klare himself is writing off the continent's population. His intentions are noble, and he is obviously making a deeper point about geostrategic calculations and energy resource competition among industrialized societies.[45] But it is important to be watchful of the political and policy implications of this kind of scenario building. Inundations will happen in

43. Ogunlade Davidson et al., "The Development and Climate Nexus: The Case of Sub-Saharan Africa," *Climate Policy* 3S1 (2003), S97–S113.

44. Michael Klare, "Avatar the Prequel: Will 'Earth's Last Stand' Sweep the 2013 Oscars?" available at www.tomdispatch.com/archive/175210/.

45. Michael Klare, *Rising Powers, Shrinking Planet: The New Geopolitics of Energy* (New York: Metropolitan, 2008).

coastal regions and islands, and they will prompt dislocations and changes. But to imagine countries effaced from the political map à la *Waterworld* or to cast them as wholly uninhabitable seems inapt. As pointed out at the outset of this chapter, this kind of conjecture has implications for the kinds of scenarios crafted for future generations, from today's grandchildren to today's grandchildren's grandchildren. And unfortunately, as discussed in chapter 3, such speculation is increasingly common among North Atlantic military and intelligence communities.

IMPACT OF CLIMATE CHANGE ON MIGRATION

To begin first with climate change, consider its impact on Africa, and then turn to migration dynamics, as this chapter does, makes sense given the line of argument. At the same time, it is important to acknowledge that the key issue of this book is human migration itself. It is the bottom line. How is climate change affecting migration flows? Is it augmenting them? Is it changing their direction and distance? That sets the stage, in turn, for examining political and policy responses. Rather than "follow the money," in this context it is more appropriate to "follow the people."

As argued, projecting the impact of a changing climate in Africa is challenging. However, historical and contemporary experiences offer insight into how people have reacted to a changing environment in the continent. Migrants move because of an array of factors in addition to environmental stress. Recent research has provided evidence that environmental change does influence migration patterns, but not in the dramatic fashion implied by imaginings of climate refugees fleeing for North Atlantic borders.[46] People reaching those borders may indeed have moved because of climate and environmental stress, but the evidence indicates that most environmentally induced migration does not, in fact, range a great distance.

Beyond the African context, briefly, the evidence for this argument is convincing. In the case of Ecuador, for example, environmental change can enhance migration out of communities, but the bulk of migration is *local*, with a *reduction* of flows to distant destinations.[47] Moreover, differences in household income and landholding affect people's abilities to migrate. Poorer people tend to migrate locally, with more affluent people being in a

---

46. Clark L. Gray, "Environmental Refugees or Economic Migrants?" (Washington, DC: Population Reference Bureau, 2010), available at www.prb.org.
47. Clark L. Gray, "Environment, Land, and Rural Out-Migration in the Southern Ecuadorian Andes," *World Development* 37: 2 (2009), 457–468.

position to contemplate international migration. As environmental degradation advances and affects people's livelihoods and incomes, their ability to migrate is sharply reduced. People in such circumstances often do not have the means to move. As climate change worsens the environment, contrary to expectations, migrants may not have the means to migrate.

Similar dynamics are evident in Nepal. Massey et al. sharply criticize the notion that environmental degradation—measured as declines in productivity and land cover and increased time required to gather firewood—leads to interregional or international migration.[48] Their finding is that people move, yes, often in anticipation of declines, but that more often than not such moves are short-distance. They find no evidence that increasing population density, declining vegetation, or declining organic inputs such as fertilizers promote departures from the local sites. Their conclusion is forceful:

> Demographers should evince considerable caution in viewing "environmental refugees" as a major component of migratory streams around the world. For the most part, environmental deterioration appears to promote local searches for organic inputs or alternative employment opportunities, not a desperate search for relief in distant lands.[49]

Arguing thusly does not imply that climate change has not and will not prompt future migrations, only that treating it as an indiscriminate threat to international society may be sharply misguided.

Even in the context of natural disasters (the Type I category in table 1.1), the evidence is compelling that migration is not as likely as commonly assumed. Where it does occur, it is largely directed toward nearby cities.[50] The devastating April 2004 tornado in north-central Bangladesh, while tragic, prompted a "non-occurrence of out-migration." The findings were similar for the horrendously destructive 2004 Indian Ocean tsunami. With annual interviews of 10,000 households in Aceh and North Sumatra, Gray demonstrates that there was displacement. However, he argues, "Most of the displaced remained in or near their origin community, a large proportion stayed with friends or family rather than entering camps, and many returned to their homes within a few months after the

48. Douglas Massey, William Axinn, and Dirgha Ghimire, *Environmental Change and Out-Migration: Evidence from Nepal* (University of Michigan Institute for Social Research: Population Studies Center Report 07-615, 2007).

49. Ibid., 22.

50. Bimal Kanti Paul, "Evidence against Disaster-Induced Migration: The 2004 Tornado in North-Central Bangladesh," *Disasters* 29: 4 (2005), 370–385.

tsunami."[51] It's true that CIM is a different concern because the degradation is chronic, but the migratory patterns associated with cataclysmic disasters are instructive.

Returning to the African context with these concerns in mind, it appears that similar dynamics have obtained. Environmental degradation, drought, and harsh conditions have long prompted migration on the continent. African history is a history of migrations. Trans-Saharan, trans-Sahelian, East African, and southern African migration systems are profoundly interconnected. Most migration in Africa stays within the continent, too, largely because the financial and cultural resources needed to migrate great distances are unavailable.[52] Moreover, the bulk of migration is local, urban, and/or within a subregion. At bottom, African migration is exceedingly complex and remains the subject of natural- and social-scientific inquiry. There are many known unknowns.

The most fundamental component to highlight is that much migration within Africa operates within "circular migration" systems formed in historical dynamics shaped by precolonial, colonial, and postcolonial globalizations. In the West African system, for example, longstanding migration patterns were reshaped by French and British colonialism, the coterminous development of cash crops for exports, and burgeoning urbanization.[53] As Findley et al. argue, during the nineteenth century French colonialism in West Africa transformed the region by introducing rural–urban migrations. It also affected gender and familial dynamics:

> The "new" migrations comprised men seeking work while their families stayed behind in the villages. Once Sahelians saw the enormous potential for such migrations, the number of labour migrants increased dramatically, particularly towards the growing cities and the plantation zones in coastal nations, and to mines after the beginning of the twentieth century.[54]

Additionally, newly established borders affected nomadic peoples—for example the Tuareg of the northern Sahel—who in turn found themselves in conflict with neighboring sedentary populations. Land confiscated for forest reserves and the commercialization of agriculture affected migration

51. Gray, "Environmental Refugees or Economic Migrants?"

52. Castles and Miller, *The Age of Migration: International Population Movements in the Modern World*, 4th ed. (New York: Palgrave Macmillan, 2010), 144.

53. Dennis D. Cordell et al., *Hoe and Wage: A Social History of a Circular Migration System in West Africa* (Boulder, CO: Westview, 1996).

54. Sally Findley et al., "Emigration from the Sahel," *International Migration* 33: 3–4 (1995), 462.

tterns. Finally, colonial enterprises such as the construction of the kar–Niger railroad in the '20s, which employed forced labor from Burkina Faso, and coastal coffee, timber, and cocoa plantations additionally transformed migration patterns within the region.

One might say that individuals make their own migrations, but they do not make them as they please. They migrate within the context of intricate familial, gender, household, generational, communal, religious, and ethnic dynamics. Of course they respond to powerful environmental and ecological signals. And these signals are part of a combination of socio-demographic factors.[55] Gender dimensions, for example, are especially important to take into account within the context of Sahelian migration and climate change, not to mention migration writ large.[56] The literature on gender offers valuable insight into the role it plays not only within individual and familial decisions but also on relations within society. In the context of West Africa, gender plays a crucial role in migration decisions, as the departure of a household member—be it a man or a woman—redounds on the family and the community.[57]

Affirming the concern that the predicted impact of climate change on Africa leads to it being written off, past considerations of the continent have often been rendered with an ecological determinism that views entire regions as a lost cause. During the droughts of the '70s and '80s, aid fatigue combined with a noxious blend of social Darwinism and Malthusian assumptions cast the human toll of the famine as unfortunate yet unavoidable. David Rain—whose name could be a clever nom de plume given his research on Sahelian drought—writes:

> After a tour of the drought-stricken region of the Sahel in 1974, then UN Secretary-General Kurt Waldheim was quoted as saying that unless aid is brought to the region fast, the countries of the Sahel region "could literally disappear." The great Sahelian famine was an inspiration in the 1970s for a subgenre of self-serving commentary about "lifeboat ethics" that asked whether starving people in overpopulated lands were deserving of our help or not. "The Sahel has

55. Sabine Henry et al., "Modeling Inter-Provincial Migration in Burkina Faso, West Africa: The Role of Socio-Demographic and Environmental Factors," *Applied Geography* 23 (2003), 115–136.

56. Lori M. Hunter and Emmanuel David, *Climate Change and Migration: Considering the Gender Dimensions POP2009-13* (University of Colorado at Boulder: Institute of Behavioral Science, 2009); and Susan Martin, *Refugee Women*, 2nd ed. (Lanham, MD: Lexington Books, 2004).

57. Geraldine Terry, "No Climate Justice without Gender Justice: An Overview of the Issues," *Gender & Development* 17: 1 (March 2009), 5–18; and Marie Monimart, *Femmes du Sahel* (Paris: Karthala, 1989).

virtually nothing to offer the rest of the world," wrote Garret Hardin, turning the region into an excuse for inaction.

Writing in the aftermath of the drought, Rain obviously disagrees with these assessments and concludes:

> Whether aid comes or not, the swath of land lying between desert and savanna in West Africa and the people who inhabit or walk across it are not likely to disappear soon.[58]

Even though moving in response to changing rainfall is a time-honored strategy, the droughts of the '70s and '80s actually led to a period of *reduced* migration within the countries of West Africa. In Ghana, for example, food prices in the wetter, southern part of the country prompted people to avoid drying areas and refrain from traveling southward.[59] Writing off the region as ultimately a sending region or casting it solely as a primary sending source for environmental refugees is clearly problematic given the fact that past drying actually *slowed* migration.

For example, Henry et al. examine the effect of rainfall time series on migration in Burkina Faso.[60] Using a sophisticated amalgamation of land degradation indicators, rainfall data, and community surveys, they find that the sharp south–north rainfall gradient *within* the country, which also demonstrates seasonal and annual variability, has a significant effect on first migration. They also demonstrate, however, that drought does not necessarily lead to more migration within the country, never mind across borders. Explanations include the provision of assistance to stricken communities and, as in the Ecuadoran context, the loss of mobility that comes with a reduction of resources. In fact, contrary to prevailing assumptions, they conclude: "Men and women from *better watered areas* are more likely to engage in a long-term migration to a foreign country than those living in the drier regions."[61]

Also of interest is the extent to which a member of a household moves to another region as part of a strategy and then moves still farther away, in

58. David Rain, *Eaters of the Dry Season: Circular Labor Migration in the West African Sahel* (Boulder, CO: Westview, 1999).

59. Kees van der Geest, "Ghana," in *EACH-FOR: Environmental Change and Forced Migration Scenarios D.3.4. Synthesis Report*, ed. Andras Vag (Brussels, Belgium: European Commission, 2009), 46–47.

60. Sabine Henry et al., "The Impact of Rainfall on the First Out-Migration: A Multi-Level Event-History Analysis in Burkina Faso," *Population and Environment* 25: 5 (May 2004).

61. Ibid., 455, emphasis added.

a "creeping onward migration."[62] This mode of movement has been documented in research on refugees fleeing conflict. The first flight to escape imminent harm leads to a period of building resources in a nearby destination, which facilitates a second, more distant, move, and so on.[63] In the context of CIM in Sahelian Africa, however, the movement is apparently often to major urban centers, then southward to coastal cities rather than northward toward the Maghreb and the Mediterranean. In the case of Niger, most of the people who migrate due to environmental problems move first to Niamey and Tilabéri.[64] Ultimately, they hope to return to the land. Some leave for Nigeria, Mali, Chad, Cameroon, and Benin for a longer period, but most return. Some leave for Europe, but they are usually affluent northerners (in Agadez) moving for prestige. If southerners leave the country, again, they move to nearby African countries. A precise accounting of the numbers of migrants remains challenging if not impossible.

Given the long-standing symbiotic relationship between pastoralists and settled farmers, in which animals are allowed to graze on plants in exchange for fertilizer and meat, stressed lands can also lead to conflict. This can be managed or exacerbated by government policy, too, as Tonah shows in his comparison of the Ivory Coast's tradition of being more receptive to Fulbe pastoralists than neighboring Ghana's.[65] As climate change affects conditions, and as settlement becomes denser, conflict could become more normal. The ongoing clashes centered around Jos, Nigeria—which erupted again in 2010—is often presented in the Western media as a religious one between Christians and Muslims, but it's also closely tied to differences between herders and settled farmers.[66]

Finally, there has been a significant greening of the Sahel since the drought of the '70s and '80s. This greening (examined, with its implications, in chapter 5) is wholly consistent with an understanding of climate change as fluctuating around a long-term signal. The intricate interrelationship

62. Koko Warner et al., *In Search of Shelter: Mapping the Effects of Climate Change on Human Migration and Displacement*, (New York: CARE International, 2009), 9.

63. Karen Jacobsen, "Refugees and Asylum Seekers in Urban Areas: A Livelihoods Perspective," *Journal of Refugee Studies* 19: 3 (2006), 373–286.

64. Tamer Afifi, "Niger," in *EACH-FOR: Environmental Change and Forced Migration Scenarios D.3.4. Synthesis Report*, ed. Andras Vag (Brussels, Belgium: European Commission, 2009), 42–44.

65. Steve Tonah, "Integration or Exclusion of Fulbe Pastoralists in West Africa: A Comparative Analysis of Interethnic Relations, State and Local Policies in Ghana and Côte d'Ivoire," *Journal of Modern African Studies* 41: 1 (2003), 91–114.

66. Sergio Tirado Herrero, "Desertification and Environmental Security: The Case of Conflicts between Farmers and Herders in the Arid Environments of the Sahel," in *Desertification in the Mediterranean Region: A Security Issue*, eds. William G. Kepner et al. (Dordrecht, The Netherlands: Springer, 2006), 109–132.

of climatic oscillations with human migratory oscillations might actually fall into Rumsfeld's third category: unknown unknowns.

CONCLUSION

As temperatures warm and precipitation becomes more erratic in portions of Sahelian and sub-Saharan Africa, the land will indeed become more stressed. In the context of the highest population growth rates in the world, traditional coping mechanisms may be challenged by a shortage of fuel wood, flooding caused by the soil's diminished absorptive capacity, food shortages, limited options for alternative livelihoods, poor governance, and ever-increasing rates of urbanization.

It is vital, however, to avoid a simplistic Malthusianism on the anticipated developments; too often, people have been seen as a threat to the environment.[67] Malthusianism informed colonial-era discourses about how indigenous populations squandered their patrimony and needed to be rescued from themselves via colonial tutelage. In the Maghreb, this took on particularly pernicious forms as French colonial administrators claimed that the local Arab and Tamazight (Berber) populations had ruined once fecund and productive lands.[68] It would be doubly tragic to replicate such ideologies in the postcolonial era.

In recent years, people motivated to move because of climate change have been increasingly cast as security threats to North Atlantic interests. Such assertions seem to ignore current knowledge about climate change, geography, and human migration patterns. Analysts see that people already live with extraordinary inequalities and harsh conditions. They conclude that climate change will alter the environment for the worse. And then they conclude that borders are going to be inundated. However, the research is persuasive that in most instances of environmental change, people move to nearby destinations as part of household strategies; they also eventually seek to return to their points of origin. Even in countries such as Senegal, which has a venerable tradition of out-migration to neighboring as well as North Atlantic countries, the evidence is strong that people moving because of rainfall challenges tend to remain within

67. Melissa Leach and James Fairhead, "Challenging Neo-Malthusian Deforestation Analyses in West Africa's Dynamic Forest Landscapes," *Population and Development Review* 26: 1 (March 2000), 17–43.

68. Diana K. Davis, *Resurrecting the Granary of Rome: Environmental History and French Colonial Expansion in North Africa* (Athens: Ohio University Press, 2007).

country.[69] Where people are displaced across borders, they will likely not range too far. For example, it is anticipated that Mozambican migrants—affected harshly by the twin pincers of drought and flood—would likely move toward South Africa.[70] Although the focus here is on the North Atlantic, it is not out of the question for South African policy makers to deploy securitized discourse in the context of an anticipated immigrant influx.

These findings suggest that policy makers instead pay more attention to cooperation between international organizations such as the IOM, the World Bank, the EU, and various development agencies and banks. Deeper and more spirited engagement with Millennium Development Goals and meeting human security needs are essential. Fair trade practices also need to be nurtured so that the dumping of low-cost commodities by North Atlantic farmers, which is ongoing, does not further undermine African agricultural sectors. Beyond this, it is essential to facilitate development by nurturing south–south cooperation on trade, sustainable development practices, and improved transportation and communication infrastructures. Such cooperation is woefully underdeveloped. For example, the absence of viable rail linkages between neighboring countries undermines trade and population movement. Even within individual countries, infrastructure must be enhanced. Ghana is roughly the size of Minnesota, yet northern Ghanaian farmers find it virtually impossible to market their produce in Accra and other southern cities because of poor transportation links.

These steps and others discussed in chapter 5 would be far better than sounding an environmental-refugee *klaxon* and securitizing international borders. Unfortunately, that is what has happened in recent years in North Atlantic and neighboring transit states.

69. Frauke Bleibaum, "Senegal," in *EACH-FOR: Environmental Change and Forced Migration Scenarios D.3.4. Synthesis Report*, ed. Andras Vag (Brussels, Belgium: European Commission, 2009), 44–45.

70. Alex de Sherbinin et al., "Casualties of Climate Change," *Scientific American* 304:1, January 2011, 67.

CHAPTER 3
The Securitization of Climate-Induced Migration

Human-induced climate and hydrologic change is likely to make many parts of the world uninhabitable, or at least uneconomic. . . . Over the course of a few decades, if not sooner, hundreds of millions of people may be compelled to relocate because of environmental pressures.

—Jeffrey Sachs[1]

The consequences of climate change will be found, and are being found now around the world. New climate conditions will drive human beings to move in ever larger numbers, seeking food, water, shelter and work. No region will be immune. Climate refugees will increasingly cross our own borders. The stress of changes in the environment will further weaken marginal states. Failing states will incubate extremism. In South Asia, the melting of Himalayan glaciers jeopardizes fresh water supplies for more than one billion human beings. . . . All of this is just the foretaste of a bitter cup from which we can expect to drink should we fail to address, urgently, the threats posed by climate change to our national security.

—Vice Admiral Lee Gunn (U.S. Navy, Retired)[2]

Climate-induced migration (CIM) is real. It has happened for millennia, of course. What's new is that in recent decades, anthropogenic climate change has joined the list of factors that impel migration. If CIM were to

1. Jeffrey Sachs, "Climate Change Refugees: As Global Warming Tightens the Availability of Water, Prepare for a Torrent of Forced Migrations," *Scientific American*, June 1, 2007. See also Jeffrey Sachs, *Common Wealth: Economics for a Crowded Planet* (New York: Penguin, 2008).
2. Lee Gunn, "Introduction," in *Climate Security Initiative: Climate Security Index* (Washington, DC: American Security Project, 2009).

remain at current levels (contributing to the "mixed flows" of migrants already endeavoring to reach North Atlantic countries), it would be a valid concern. But the evidence is compelling that in the years to come climate change will only deepen. As chapter 2 sought to demonstrate, how this may affect future CIM remains unclear.

The impact of climate change has been and will be distinctive in different regions of the world. Regions between the equator and the descending portions of the Hadley circulation cells associated with 30° latitude north and south are home to the world's great deserts and are expected to experience enhanced drying. Low-lying islands and coastal communities in the Indian and the Pacific oceans will experience more frequent flooding and the salinization of groundwater. People in South Asia, especially in Bangladesh and Myanmar, will experience ever worse inundations and the disappearance of livable land. Humans are innovative creatures who adapt to a broad spectrum of inhospitable climates, and societies often remain in place in harsh circumstances. Movement is also a form of adaptation, however, so it is reasonable to conclude that migration might increase. Migratory flows will continue to be a challenge for the governments of sending states, receiving states, and transit states. The politics of transit states are treated in chapter 4. This chapter grapples with the response of North Atlantic receiving countries, principally the United States and the European Union.

If there is good news to be had, it is that governments are beginning to take climate change seriously. With varying degrees of commitment, EU member countries have done so for years, stretching back to the 1992 United Nations Conference on Environment and Development (UNCED, known as the Earth Summit or the Rio Summit). The Rio Summit brokered the UN Framework Convention on Climate Change (UNFCCC), which led to the signing in 1997 of the Kyoto Protocol. Copenhagen in December 2009 was the 15th United Nations Climate Change Conference (COP15). Throughout the last two decades, the EU has been aggressive in its support for the Kyoto Protocol, and member countries are often leaders in technological innovation on energy policy and mitigation efforts—as Holland, Denmark, and Germany demonstrate with respect to alternative forms of energy.

As for the United States, during the '90s environmental advocates typically viewed the Clinton-Gore administration's efforts with respect to the environment as dismayingly insufficient. This criticism lasted until January 20, 2001, and the inauguration of the Bush-Cheney administration. In June 2001, the United States formally withdrew from the Kyoto Protocol. As discussed below, the Bush-Cheney administration's opposition to environmental causes and, specifically, climate change mitigation began to soften somewhat in its second term, although this is more likely associated with pressures from a Democratic-controlled Congress after the November 2006 elections. Moreover, public consciousness began to change by 2007, in part because of the release of former Vice President Al Gore's *An Inconvenient Truth* and its receipt of an Oscar for best documentary film, and the joint awarding of a Nobel Peace Prize to Gore and the Intergovernmental Panel on Climate Change (IPCC). The IPCC also published its Fourth Assessment Report in November 2007, following the release of the UK Government's Stern Report in late 2006. As Dabelko notes, using a droll euphemism given the precipitous decline in the home delivery of newspapers, "Climate change has become an above-the-fold issue in the last few years."[3]

"Climate skeptics" and "climate deniers" persevere, of course. Their political heft is significant. In some instances, their activities can engender clumsy politics and counter-hyperbole on the part of climate scientists, as demonstrated by the November 2009 hacked e-mails from the University of East Anglia's Climate Research Unit. What is evident, though, is that despite the noise made by skeptics and deniers, governments and policy makers have stepped up efforts to address climate change issues. More than that, conventional political, military, and economic institutions are emphasizing "energy security" and other notions of the security challenges posed by climate change. U.S. media personalities like Rush Limbaugh and Glenn Beck—along with Senator Inhofe and House Republicans after the 2010 election—may think climate change is a hoax, but U.S. military and intelligence officials certainly do not.

This chapter argues that not only has the discourse associated with climate change taken a decidedly security-oriented turn, but so has the approach to CIM.[4] Moreover, it is a discursive and empirical turn that is successful politically. In its more strident forms, this "securitization" emphasizes military preparedness and support for harsh, vigorous border control measures. It also entails shoring up transit states on North Atlantic borders, even if it means enhancing their authoritarian character. In its less strident forms, it calls for greater attention to intelligence and collaboration on security between states. In either case, the emphasis is on the potential for conflict stemming from CIM and the concomitant need for

3. Geoffrey Dabelko, "Planning for Climate Change: The Security Community's Precautionary Principle," *Climate Change* 96 (2009), 13–21.

4. Betsy Hartmann, "Rethinking Climate Refugees and Climate Conflict: Rhetoric, Reality and the Politics of Policy Discourse," *Journal of International Development* 22 (2010), 233–246.

the state to respond vigorously. In media coverage and electoral calculations, this strategy works well, as voters find such discourse compelling.

It is true that some parts of the state apparatus are more inclined to focus on mitigation efforts. As Allison instructs in his analysis of the Cuban missile crisis, bureaucratic interests can play crucial roles within a government. Where you stand on policy matters depends on where you sit in the bureaucracy.[5] Environmental or energy ministries, as well as ministries devoted to labor, industry, and agriculture, are taking climate change seriously. They are working to craft mitigation efforts, albeit bearing in mind the interests of respective constituencies, and albeit perhaps not as quickly or as sufficiently as climate scientists recommend. For their part, by contrast, intelligence and security bureaucracies, including defense and interior ministries, are working assiduously on the anticipated *threat* posed by climate-induced migration. As Dabelko points out, "[This] should not be a surprise. Security actors, like the insurance industry, are in the game of assessing risk, and the message coming from scientists is that climate change poses many hazards."[6] Security officials would not be doing their jobs if they were not examining prospective risks.

What is crucial is that it is the security efforts—as opposed to those devoted to mitigation—that receive the most political support from electorates. North Atlantic citizens' sense of security is reassured when they perceive governments to be tough on immigration and on stopping unwanted flows of migrants.[7] Security is perceived as an efficacious, low-cost way of "doing something," "being tough," and maintaining stability in a changing, uncertain world. Mitigation efforts at the national or even subnational (state or city) level, on the other hand, are more challenging politically and can be perceived as costly economically. For politicians, support of mitigation efforts can be devastating at the polls. (Ironically, the cost to the taxpayer of military expenditures is exorbitant. By one estimate, it is 42 cents on every dollar in the United States.[8] While a fraction of that goes to immigration control, the fraction may very well exceed the "cost" of altering one's behavior.)

5. Graham Allison and Philip Zelikow, *Essence of Decision: Explaining the Cuban Missile Crisis*, 2nd ed. (New York: Longman, 1999).

6. Dabelko, "Planning for Climate Change: The Security Community's Precautionary Principle."

7. Martin O. Heisler and Zig Layton-Henry, "Migration and the Links between Social and Societal Security," in *Identity, Migration and the New Security Agenda in Europe*, eds. Ole Wæver et al. (New York: St. Martin's, 1993), 148–166; and Mark B. Salter, "Passports, Mobility, and Security: How Smart Can the Border Be?" *International Studies Perspectives* 5 (2004), 71–91.

8. See National Priorities Project at www.nationalpriorities.org.

Here is the deeper point: "climate refugees" are an easily invoked specter that ties into a citizenry's deepest fears about climate change. One might become resigned to the inevitability of warmer temperatures, heat waves, harsher precipitation patterns, floods, droughts, pests, blighted crops, and so on. One might even accept adaptation as unavoidable. But droves of invaders? Immigrants and refugees streaming toward "our" border? That fear is hard to bear, and it is easily mobilized by media-savvy politicians, as well as by the media's panic entrepreneurs. At best, one might have a caring, charitable view toward people seeking to cross forbidding borders. Care-oriented NGOs and religious groups often display this impulse in the American Southwest or in the Mediterranean basin. More common, however, is the view that outsiders are a threat. They need to be kept out at all costs.

This chapter proceeds first by exploring the evolution of the discourse associated with environmental security and environmental conflict. It does so by examining definitions pertaining to security and environmental security. It then turns to the increasing injection of security imperatives into climate change discourse, overlaying such developments with the coterminous securitization of immigration politics, especially after September 11. Finally, it treats at length the thorny politics associated with CIM.

SECURITY, ENVIRONMENTAL SECURITY, AND ENVIRONMENTAL CONFLICT

How does one understand the powerful notion of security? Invoked in so many contexts and operating on so many levels, *security* is a master noun of political discourse. In political philosophy and theories about international relations, state security is at the heart of Thucydides' "Melian Dialogue" from the *History of the Peloponnesian Wars*, Machiavelli's *The Prince*, and Hobbes's *Leviathan*. Nevertheless, Levy's trenchant 1995 critique of the environmental security literature remains relevant.[9] He argues that, for all the preoccupation with environmental security, scholars rarely stop to offer basic definitions.

Definitions are dicey, as was evident in the IOM definition of environmental migrants. Beginning an analysis with "According to Webster's . . ." risks fixing the meaning of a word in a static context, whereas its meaning evolves depending on discursive context and dialectical relationships with

9. Marc Levy, "Is the Environment a National Security Issue?" *International Security* 20: 2 (Fall 1995), 35–62.

_he point with *security* is precisely that: its meaning is dynamic. As
ᵼ be evident, in the last several decades its meaning has evolved considerably. Nonetheless, to begin, this analysis defines security as *the condition of
being protected from threat and avoiding the absence of anxiety stemming from
a fear of perceived danger*. Being secure, then, can operate on many levels:
individual, familial, communal, national. Maintaining security also requires
a vigilance that must strike a balance between complete security and an
undesirable modus vivendi. After all, true security might mean locking oneself in a room with a stockpile of H_2O and Twizzlers. Perhaps one would
also have something to read, DVDs of *The Office*, and a fully loaded iPod. Yet
it would hardly constitute "positive freedom" or a life worth living.[10]

If all pursue an extreme form of security, it can lead to greater insecurity. And individual pursuit of security can prompt harm. In *Once upon a
Time*, a breathtaking short story by South African Nobel Prize winner
Nadine Gordimer, a husband and wife escalate the security fencing around
their villa.[11] Each new precaution is deemed insufficient. Initially, they
have a reasonable wall around their property, but in their neighborhood
they see other walls topped with shards of glass and spikes. Ultimately,
responding to an advert for "Total Security," the couple decide to install
curled razor wire. At last, they *feel* secure . . . until their toddler is savagely
injured when he climbs into the wire as part of an imaginary mission to
save a sleeping princess. Seemingly sensible precautions by an individual—
a gun in an upstairs drawer—can lead to tragedy.

The traditional texts in the twentieth-century field of international relations tend to offer definitions based on vague and circular reasoning. Prior
to the end of the Cold War in the late '80s, security was invariably crafted
as military in orientation, preoccupied with defending a national territory
from the external threat of a conventional or nuclear military attack.[12] The
resultant "security dilemma," first articulated by Herz in 1950, occurs
because one state's military preparation is perceived as a threat to other
states. He writes:

> Striving to attain security from . . . attack, [states] are driven to acquire more
> and more power in order to escape the impact of the power of others. This, in

---

10. Amartya Sen, *Development as Freedom* (New York: Oxford University Press, 1999).
11. Nadine Gordimer, "Once upon a Time," in *Jump and Other Short Stories* (New
York: Penguin, 1992).
12. Hans Morganthau, *Politics among Nations: The Struggle for Power and Peace* (New
York: Knopf, 1960); and Stephen Walt, "The Renaissance of Security Studies," *International Studies Quarterly* 35: 2 (1991).

turn, renders the others more insecure and compels them to prepare for the worst. Since none can ever feel entirely secure in such a world of competing units, power competition ensues, and the vicious circle of security and power accumulation is on.[13]

In a state system characterized by anarchy, it is necessary to pursue defensive preparation and the construction of a balance of power to thwart the emergence of an irresponsible and coercive hegemonic power. However, this can make matters worse, as arms races between gangs, or countries, illustrate.

In addition to being imprecise and inclined toward tautology—security is the absence of a military threat to be secured by pursuing military preparedness, which can heighten insecurity—the traditional military notion of security was insufficient when considering environmental, epidemiological, and other potential threats to a country. Writing before the end of the Cold War, but anticipating the evolution of the concept, Ullman offered a valuable understanding:

> A threat to national security is an action or a sequence of events that (1) threatens drastically and over a relatively brief period of time to degrade the quality of life for the inhabitants of a state, or (2) threatens significantly to narrow the range of policy choices available to a state or to private, non-governmental entities (persons, groups, corporations) within the state.[14]

This framing allows a better understanding of different conceptions of security beyond *strategic security* preoccupied with military threats. For example, unease with *economic security* became more prominent in the aftermath of the oil shocks in the '70s, the perceived decline of U.S. hegemony, and the threat to U.S. hegemonic stability.[15] This also encompasses *societal security*, the emphasis of the Copenhagen School on the perception that a society's distinct cultural and historical identity is being threatened by outside forces.[16] The pursuit of "security" in this context is

---

13. Quoted in John Mearsheimer, *The Tragedy of Great Power Politics* (New York: Norton, 2001), 36.

14. Quoted in Levy, "Is the Environment a National Security Issue?" 40.

15. Charles Kindleberger, *The World in Depression, 1929–39* (Berkeley: University of California Press, 1973); and David A. Lake, "British and American Hegemony Compared: Lessons for the Current Era of Decline," in *International Political Economy: Perspectives on Global Power and Wealth*, eds. Jeffry Frieden and David A. Lake, 3rd. ed. (New York: St. Martin's, 2000), 127–140.

16. Michael C. Williams, "Words, Images, Enemies: Securitization and International Politics," *International Studies Quarterly* 47 (2003), 511–531.

a constructed process that reinforces the sense of self as distinct from an Other.[17]

Common to these different dimensions is the use of security as a discourse, a speech act articulated by an authority and heard by an audience that is, presumably, unaware of the risk. Articulating something as a matter of security, therefore, involves convincing oneself and an audience that a threat exists and that it must be met. In its demagogic form, security is pitched as something understood as necessary only by the authority, who must open the ears and eyes of others: "Am I the only one who sees what's going on here! Listen. It's like this . . ." This is parodied in Stanley Kubrick's *Dr. Strangelove*, when General Jack D. Ripper seeks, in a conspiratorial whisper, to convince Group Captain Lionel Mandrake that Communists are fluoridating the water.

In subtler and less obsessive forms, security discourse is often pitched as a sober assessment of likely risks, with assertions that the threats will be met. Securitization is an active process, then: one that identifies a threat, specifies its character, taps into a "social imaginary" of fear, and crafts a response that, presumably, is robust and effective in enhancing safety.[18] How an authority or responsible power responds to a threat is crucial. For example, whether a disease is identified as a threat to the nation shapes whether doctors, a Ministry of Health, or security officials are the primary responders. Elbe points to this concern in his analysis of the securitization of HIV/AIDS, illustrating that a "threat-defense" logic might both hamper international cooperation because of the nationalization of the public health response and perpetuate stereotypes against people living with HIV/AIDS.[19] When a swine flu epidemic emerged in Mexico in April 2009, the Department of Homeland Security Secretary Janet Napolitano was the government point person, not the head of the Centers for Disease Control or the American Medical Association.

What, then, to make of *environmental security*? In the aftermath of the Cold War and the turn to an anticipated "New World Order," scholars and policy makers increasingly pointed to the close connection between environmental concerns and national security. For example, some scholars offered alarms about the potential for environmental scarcity to prompt conflict. To

17. Gregory White, "Sovereignty and International Labor Migration: The 'Security Mentality' in Spanish-Moroccan Relations as an Assertion of Sovereignty," *Review of International Political Economy* 14: 4 (2007), 690–718.

18. Simon Dalby, *Security and Environmental Change* (Malden, MA: Polity, 2009).

19. Stefan Elbe, "Should HIV/AIDS Be Securitized? The Ethical Dilemmas of Linking HIV/AIDS and Security," *International Studies Quarterly* 50: 1 (March 2006), 119–144.

be sure, this was not entirely a post–Cold War phenomenon. The UN World Commission on Environment and Development, known as the Brundtland Commission, had anticipated the discursive shift in 1987 in arguing, "The whole notion of security as traditionally understood—in terms of political and military threats to national sovereignty—must be expanded to include the growing impact of environmental stress—locally, nationally, regionally and globally."[20] Barnett points out that the climate change–security discourse emerged as early as 1971, with Richard Falk's *This Endangered Planet*.[21]

Depending on a policy maker's political position or a scholar's research ambit, framing environmental issues as a security issue is often cast as a means of getting attention from policy makers. After all, if it is a security concern, it is a matter of "high politics" that requires immediate and full response by the state. In 1992, then Senator Al Gore's *Earth in the Balance* made the case that environmental matters should be taken more seriously as a security question.[22] In his introduction to a 1994 edition of Rachel Carson's seminal *Silent Spring*, then Vice President Gore proclaimed environmental matters to be an appropriately central concern of government.[23]

As Deudney, Levy, and others warned in the '90s, however, the problem with more extreme forms of this effort to cast the environment as a security threat is that they fail to distinguish which concerns are genuine threats. Further, the state's powerful security apparatuses might be the last things that environmentalists would want brought to bear on environmental issues. Given the high ecological costs of military activity—not to mention the military's penchant for secrecy—security apparatuses would likely do more harm than good and would be ultimately unaccountable.[24] It's a closely guarded secret where the combat tank is idling and belching emissions. And the EPA has no effective oversight on an air force base's disposal of toxic materials. Similar critiques have emerged from feminist scholarship. Tickner, for example, challenges the realist and masculine assumptions inherent in approaches to environmental security.[25]

20. United Nations World Commission on Environment and Development, *Our Common Future* (New York: United Nations, 1987).

21. Jon Barnett, "Security and Climate Change," *Global Environmental Change* 13 (2003), 7–17.

22. Al Gore, *Earth in the Balance* (New York: Houghton Mifflin, 1992).

23. Al Gore, "Introduction," in Rachel Carson, *Silent Spring* (New York: Houghton Mifflin, 1994).

24. Daniel Deudney, "The Case Against Linking Environmental Degradation and National Security," *Millennium* 19: 3 (1900), 461–476.

25. Jill Tickner, *Gender in International Relations* (New York: Columbia University Press, 1992). See also Jody Seager, *Earth Follies: Coming to Feminist Terms with the Global Environmental Crisis* (New York: Routledge, 1993).

By the mid-'90s, environmental security was being treated more broadly, as more than a matter of military force and national security. A less statist orientation emerged with the 1994 United Nations Development Program report, which outlined different dimensions of security: economic, food, health, political, personal, and so on. In many ways, as with the Brundtland Commission's work, it was an appealing discourse because it emphasized the challenges of environmental degradation for all human beings. It also stressed the importance of lessening high consumption patterns, reducing population growth, and so on. Finally, it looked out on long-range solutions, rather than short-term fixes. It was much less state-centric, devoted more to societal and individual levels in searching for security.

Whether this risks making security so all-encompassing as to be meaningless is open to debate. One might argue that if everything is a security threat, then *security* becomes useless as an analytic category. Some suggest, however, that if coupled with a specification of the risk at hand, the frame of environmental security can help move debates beyond narrow, state-centric concerns. What is striking in recent years, in contrast to the '90s (as treated more systematically below), is the degree to which environmental issues have become accepted as security concerns.

Since September 11, it is almost as if mere environmental security has become inadequate as a means of mobilizing policy makers and electorates. Instead, the fixation is on environmental *conflict*, which goes a step deeper into the environmental security rubric. The discourse centers on "the possibility that groups within society will engage in violent conflict as natural resource stocks diminish due to environmental degradation."[26] This framework often looks to the central role of the state and its security apparatus. In many instances, the state's provision of security can mean a narrower security—security for the state only, not for societal actors.

In the 20 years since the end of the Cold War, there has been an intriguing evolution concerning climate change. Initially, the preoccupation with scarcity and environmental issues prompting conflict focused on issues such as fisheries, ozone depletion, and loss of agricultural land and forests.[27] By the mid-'00s, however, scholars and, increasingly, security officials were examining the connections between climate change and security.[28]

---

26. Nicole Detraz and Michele Betsill, "Climate Change and Environmental Security: For Whom the Discourse Shifts," *International Studies Perspectives* 10 (2009), 305.

27. Thomas Homer-Dixon, "Environmental Scarcities and Violent Conflict: Evidence from Cases," *International Security* 19: 1 (1994), 5–40.

28. Ragnhild Nordås and Nils Petter Gleditsch, "Climate Change and Conflict," *Political Geography* 26: 26 (2007), 627–638; and Jon Barnett and W. Neil Adger, "Climate Change, Human Security and Violent Conflict," *Political Geography* 26 (2007), 639–655.

Prominent social scientists like Jeffrey Sachs and Thomas Homer-Dixon emphasized that climate change would become a "threat multiplier."

Still, in considering the intriguing ways in which climate-induced migration knits together environmental conflict and the securitization of immigration politics, it is important to exercise caution. As Dabelko warns policy makers, one should not oversell the connections between climate change and violent conflict. Echoing Deudney's concerns, Dabelko argues that climate change, poverty, development, and resource use are so complicated that ham-handed security-minded responses might only exacerbate matters.[29] As Haas further argues, an environmental conflict perspective can be an easy harbor for Malthusian inclinations and thereby unwittingly support a strong state response.[30] Thus, to examine the emergence of a robust, securitized approach on the part of North Atlantic countries and transit states is not to suggest that it is a desirable norm. To the contrary, the argument here is that wearing security glasses can make one shortsighted.

THE SECURITIZATION OF IMMIGRATION

What is striking is that the deepening emphasis on environmental security has evolved in complementary ways with the securitization of immigration politics. It would be conventional to use September 11 as a pivot point in the preoccupation with security, yet it would be inappropriate to treat September 10 as unconcerned with security. Indeed, on both sides of the Atlantic Ocean, the security dimension with respect to immigration emerged fully in the '80s.[31] As Guild argues, "What happens is that foreigners, described in various different ways (migrant, refugee, etc.), become caught in a continuum of insecurity. . . . Many insecurity discourses are promoted at any given time . . . [yet] the ease with which the category of the foreigner may be added to an insecurity discourse, with the effect of heightening the perceived seriousness of the threat, remains constant."[32]

29. Geoffrey Dabelko, "Avoid Hyperbole, Oversimplification When Climate and Security Meet," *Bulletin of the Atomic Scientists*, August 24, 2009.

30. Peter Haas, "Constructing Environmental Conflicts from Resource Scarcity," *Global Environmental Politics* 2: 1 (February 2002), 1–11.

31. Myron Weiner, *Global Migration Crisis: Challenges to States and to Human Rights* (New York: Harper Collins, 1995); Peter Andreas and Timothy Snyder, eds., *The Wall around the West: State Borders and Immigration Control in North America and Europe* (Lanham, MD: Rowman & Littlefield, 2000).

32. Elspeth Guild, *Security and Migration in the 21st Century* (Malden, MA: Polity Press, 2009).

For Europe, such securitization was tied into the waning of the Cold War, deepening European integration, and the movement toward free internal movement of European citizens. If there is a birth date for Europeans' security focus on immigration, it might be June 14, 1985, when the Schengen agreement was signed. Five of the signatories of the 1957 Treaty of Rome signed it: Belgium, the Netherlands, West Germany, France, and the host, Luxembourg. The sixth original member, Italy, did not, nor did the other EEC members at the time: the United Kingdom, Greece, or Denmark. Generally speaking, at the time, immigrants did not present security concerns per se for European officials. Right-wing politicians such as Jean-Marie Le Pen, leader of France's National Front, were certainly in full force and were xenophobic in their politics.[33] Of course, nativist and anti-immigrant sentiment has been evident throughout history. In the nineteenth century, Chinese immigrants to the United States were cast as "pollution," an overlapping of discourses that anthropologists such as Mary Douglas have examined well.[34] Despite their prominence, the electoral success of the anti-immigrant right remained relatively subdued throughout the Cold War. Spain, for its part, did not even possess an anti-immigrant party until the Popular Party (PP) electoralized the issue in the '90s. Security-oriented efforts toward immigrants, to the extent they existed, were small-scale and largely a matter for respective interior ministries.

This began to change in the late '80s and early '90s. European governments working at the supranational level further solidified efforts to facilitate the free internal movement of citizens with the passage of the Single Europe Act of 1987, the deepening of Schengen in 1990, and the 1992 signing of the Maastricht Treaty to establish the EU. By extension, third-country nationals—non-Europeans—had to be put under control and treated ipso facto as security concerns. During the early '90s, European preoccupations with societal security tipped immigration away from economic concerns to a greater anxiety about incomers threatening the European way of life. By the time the 1995 Barcelona Declaration established the Euro-Mediterranean Partnership, immigration

33. Gary Freeman, "Immigrant Labour and Working-Class Politics: The French and British Experience," *Comparative Politics* 11: 1 (1978), 25–41; and Hans-Georg Betz, "The New Politics of Resentment: Radical Right-Wing Political Parties in Western Europe," *Comparative Politics* 26: 4 (1993), 413–427.
34. J. David Cisneros, "Contaminated Communities: The Metaphor of 'Immigrant as Polution' in Media Representations of Immigration," *Rhetoric and Public Affairs* 11: 4 (2008), 569–602.

was used as a full foil for Europe's sense of essential identity and citizenship.[35]

Analogous developments occurred in the United States, although it was driven less by a sense of citizenship. Throughout the '80s, anti-immigration discourse was real, but it was usually put in terms of economic security. The 1986 Immigration Reform and Control Act (IRCA, or the Simpson-Mazzoli Act) emphasized enforcement of workplace identity checks, but it was not fully framed as an effort to strengthen national security or protect America's essential identity. Ironies abounded during the 2008 Republican primaries, when candidates met for a debate at the Ronald Reagan Presidential Library and claimed Reagan's mantle as an anti-immigrant politician. Paradoxically, Reagan himself had eschewed an anti-immigrant approach and during the '80s was criticized from the right as too willing to offer amnesty to immigrants. Indeed, it was only with the movement toward the signing of the North American Free Trade Agreement (NAFTA) in December 1992, the same year as the Maastricht Treaty, that the U.S.-Mexican border became more fully securitized.[36] Shortly thereafter, in 1994, the Clinton administration began Operation Gatekeeper, marking the full emergence of a security discourse that deepened further with the 1996 Illegal Immigration Reform and Immigrant Responsibility Act (IIRIRA).

By the late '90s, then, North Atlantic countries viewed immigrants as part of a basket of security issues. Most of the rhetoric was framed in terms of societal security, but it also took the form of concern for national security. In Europe, this was fueled by terrorist attacks. The 1995 bombing in Paris's St. Michel metro station is a notorious example, as Algerian extremists extended the raging civil war in Algeria into the former *métropole*. Certainly, after September 11 and subsequent attacks in Madrid on March 11, 2004, and in London on July 7, 2005, the connection was fully made. It is crucial to note that while the Madrid bombing was perpetrated by immigrants, the London bombing was not.

Throughout the '00s, immigration policy was consistently securitized. In the United States, for example, the new Department of Homeland Security subsumed the Immigration and Naturalization Service

---

35. Bichara Khader, "Immigration and the Euro-Mediterranean Partnership," in *The Euro-Mediterranean Partnership: Assessing the First Decade*, eds. Haizam Amirah Fernández and Richard Youngs (Barcelona: Fundación para las realaciones internacionales y el dialogo exterior, 2005), 83–92.

36. Andreas, "Redrawing the Line: Borders and Security in the Twenty-First Century"; and Hufbauer and Vega-Cánovas, "Whither NAFTA: A Common Frontier?"

(INS), which became Immigration and Customs Enforcement (ICE). In terms of discourse, literature flourished suggesting that immigration must be stopped because, as Krikorian argued, terrorists would "come here and kill our children in their beds."[37] While preoccupied with a broader sense of societal security and national culture, Huntington's claim that Mexicans and other Latinos would undermine America's unique Anglo-Protestant values and political culture emerged in the post-9/11 context, too.[38]

In Europe, post-9/11 anxieties have not died down at all, as is evident not only in politics but also in the broader culture. It has been fueled by the appearance of what Lalami calls "the alarmist tract . . . a whole genre of non-fiction."[39] Titles reveal much about the perceived threat confronting Europe: Oriana Fallaci's *The Rage and the Pride* (2002), Bat Ye'Or's *Eurabia: The Euro-Arab Axis* (2005), Melanie Phillips's *Londonistan* (2006), Claire Berlinski's *Menace in Europe: Why the Continent's Crisis Is America's Too* (2006), Bruce Bawer's *While Europe Slept: How Radical Islam Is Destroying the West from Within* (2006) and *Surrender: Appeasing Islam, Sacrificing Freedom* (2009), and Christopher Caldwell's *Reflections on the Revolution in Europe: Immigration, Islam, and the West* (2009). In an interview about Caldwell's book, Christopher Dickey, the Paris bureau chief of *Newsweek* who has lived in Europe for decades, pronounced the argument that Europe will somehow be overwhelmed by Muslims as "nuts."[40] Yet there is no denying the popularity of the genre as well as the influence such a perspective has had on public opinion and on policy orientations.

The bottom line is that in recent decades there has been a striking amalgamation of security imperatives, with immigration nestled at their intersection. Nonetheless, until the recent emergence of CIM as a security concern, environmental security and immigration security operated on separate tracks. To be sure, previously, there were instances in which environmental discourse was used against immigrants. As illustrated by battles within the Sierra Club, anti-immigration politics were sometimes

37. Tamar Jacoby and Mark Krikorian, "Debates: Dealing with Illegal Immigrants Should Be a Top Priority of the War on Terror," *National Review Online*, February 12, 2003, available at www.nationalreview.com.

38. Samuel Huntington, *Who Are We? The Challenges to America's National Identity* (New York: Simon & Schuster, 2005).

39. Laila Lalami, "The New Inquisition," *The Nation*, December 19, 2009, available at www.thenation.com.

40. National Public Radio's "On Point," August 12, 2009, available at www.onpointradio.org/2009/08/europe-and-islam.

wrapped in a green cloth.[41] The argument here, however, is that with the emergence of CIM as a perceived threat, environmental security has now fully joined with immigration security as an ostensible cause for alarm. CIM is the seamless dovetail of immigration security and environmental security discourses.

THE SECURITIZATION OF CLIMATE-INDUCED MIGRATION (CIM)

As examined in chapter 1, the potential for CIM to fuel a threat-fear response and contribute to a securitization imperative among states was recognized early on. Shortly after El-Hinnawi and others wrote about climate refugees in the mid-'80s, and Myers amplified the scenario, scholars pointed out not only that there were empirical challenges but also that CIM could be cast as a security threat.[42] A statist, securitized logic would proceed as follows:

- If protection for political refugees is difficult to secure;
- If there is little to no absorptive capacity for economic migrants in a national economy;
- If cultural differences between citizens and noncitizens are too broad; and/or
- If newcomers pose criminal or terrorist threats because of political orientation;
- Then governments will be obliged to guard against the threat of incoming people.

The caution of such analysts is absolutely central to the logic and preoccupation of this book. Governments would need only point to CIM as a real and growing phenomenon to be off the hook, relieved (along with allies that support them) of responsibility for solving other problems that cause population movements.[43]

41. Leslie King, "Ideology, Strategy and Conflict in a Social Movement Organization: The Sierra Club Immigration Wars," *Mobilization: An International Quarterly* 13: 1 (2008), 45–61.

42. Gaim Kibreab, "Environmental Causes and Consequences of Migration: A Search for the Meaning of 'Environmental Refugee,'" *Disasters* 21: 1 (1997), 20–38; and Stephen Castles, *Environmental Change and Forced Migration: Making Sense of the Debate* (Geneva, Switzerland: UNHCR Evaluation and Policy Analysis Unit, 2002).

43. Idean Salehyan, "The New Myth about Climate Change," *Foreign Policy*, August 2007, available at www.foreignpolicy.com.

Arguing for caution, however, is not to deny that CIM is happening in certain parts of the world. One should not criticize the concept because of its potential political uses. A city might offer a feeble or, perhaps, counterproductive response to homelessness; it might build overnight shelters to the exclusion of affordable and transitional housing, and social services such as needle exchange or drug and occupational counseling might be wholly inadequate. Yet that does not mean homelessness is not happening. Similarly, security actors and governments may use CIM in inappropriate and counterproductive ways, but that does not mean that CIM does not merit analysis.

To be sure, the real-world experience of migrants storming fences, invading in flotillas of boats, or being smuggled in containers—and the threat of more to come—continues to make the security logic appealing. Media sources in the North Atlantic often depict such images: migrants crossing bleak deserts in Arizona or Algeria, breaching fences in Texas or Ceuta, lining the gunwales of intercepted boats off the coast of Tarifa or Lampedusa or Sicily. The international media's attention to the "storming" of the fences in Ceuta and Melilla in 2005 (examined closely in chapter 4) shied away from analyses of Spanish and Moroccan authorities' draconian responses.[44] It also neglected to examine the two governments' close collaboration. Instead, it appealed to its varied audiences' assumptions that migrants are consumed with sheer desperation to get into Spain.

Throughout the '90s and the early '00s, even with right-wing governments keen on a security discourse seeking to emphasize immigration fears, CIM simply did not gain traction as a concern. The Bush administration and conservative European leaders such as Spain's José María Aznar, Italy's Silvio Berlusconi, France's Jacques Chirac (and then Interior Minister Nicolas Sarkozy), and Portugal's José Manuel Durão Barroso did not explicitly single out CIM. Nor did center or center-left leaders such as Germany's Gerhard Schröder, Canada's Jean Chrétien and Paul Martin, Spain's José Luis Rodríguez Zapatero, or England's Tony Blair. Why? While climate change was accepted by the scientific community as an empirical phenomenon, as evidenced in the 2001 Third Assessment Report of the IPCC, its understanding and acceptance by officials and the broader public was, at best, incomplete. By the IPCC's AR4 in 2007, and certainly by COP15 in Copenhagen in 2009, broader acceptance of the science of climate change

---

44. In Moroccan Arabic these territories are known as Sebta and Melillia. On occasion, depending on the context, the names are used interchangeably. More often, however, Ceuta is preferred as Spain has official sovereignty.

was in place; what had intensified was the debate about the appropriate policy response to it.

If the impulse to securitize immigration emerged in Schengen in 1985, and in the United States in the late '80s as the inevitability of NAFTA became apparent, when did the securitization of CIM begin? Certainly, CIM received attention from analysts, scholars, and NGOs throughout the '90s and into the '00s, as discussed in chapter 1. International organizations like the UNHCR included climate refugees as a concern, albeit it with the now familiar questions about the "refugee vs. migrant" label and the contexts and responses. But when did government officials become cognizant of CIM's political salience, its potential as a threat to security, and the need to offer, or at least plan for, a robust response?

BEYOND THE NORTH ATLANTIC—SOUTH ASIA, THE PACIFIC, AND SOUTHERN AFRICA

An answer to this question might be found beyond the North Atlantic context. For example, one sees Indian and Bangladeshi officials in the early '00s expressing deep anxieties about Bangladesh's vulnerability to flooding. In 2003, India began construction of a 2,100-mile high-tech "separation barrier."[45] The officially stated (and ostensibly plausible) fear was of infiltration by Bangladeshi Islamists. Completed in early 2010, the fence has been fraught with tragic politics on the ground: border communities divided, property claims fenced, homes lost, and informal trade disrupted.[46] India's efforts were supported by Washington and NATO, even as Bangladesh has also been a reliable geostrategic ally. Nevertheless, as Bangladesh's vulnerability to climate change and its prospects for greater flooding have become more worrisome, being surrounded by razor wire has reified and hardened a precarious border—with Indian officials increasingly inclined to cite climate refugees, rather than the Islamist threat, as a concern.[47] Is this a calculated decision to deploy CIM because it might be more persuasive as a rationale? Or because the Islamist threat has become

45. Ali Riaz, "Bangladesh," in *Climate Change and National Security: A Country-Level Analysis*, ed. Daniel Moran (Washington, DC: Georgetown University Press, 2010), 103–114.

46. Reece Jones, "Geopolitical Boundary Narratives, the Global War on Terror and Border Fencing in India," *Transactions of the Institute of British Geographers* 34 (2009), 290–304.

47. Lisa Friedman, "How Will Climate Refugees Impact National Security?" *Scientific American*, March 23, 2009.

less politically expedient? It is impossible to know. One final note about Bangladesh: the only place where it does not border India is in the southeast, where it shares a 500-mile frontier with Burma/Myanmar. There is no fence there, yet one would be hard-pressed to argue that Bangladeshi-Myanmar relations are peaceable and without controversy. Moreover, should there be migration pressures from Bangladesh, the construction of the fence on the Indian-Bangladeshi border shifts responsibility for absorption onto Myanmar.

At the subnational level in South Asia, migration within India is also a concern. In western India, for example, Gujarat and Rajasthan are expected to experience water scarcity and enhanced urbanization. And storm surges in major coastal urban areas such as Mumbai and Kolkata can be expected to induce migration away from the sea.[48] Indian officials also express concerns that Himalayan glaciers will melt rapidly, initially causing flooding and then drought. Melting has occurred in recent decades and is anticipated to accelerate. That said, in February 2010 the pace of the Himalayan melt was at the heart of a controversy associated with the accuracy of the IPCC estimates. The Himalayan glaciers—aka the "Water Tower of Asia"—feed the Indus River, the Brahmaputra, the Mekong (which descends into Southeast Asia), the Irrawaddy in Myanmar, and the Yellow and Yangtze rivers of China. Retired Air Marshal A. K. Singh, a former commander in India's air force, foresees mass migrations:

> It will initially be people fighting for food and shelter. When the migration starts, every state would want to stop the migrations from happening. Eventually, it would have to become a military conflict. Which other means do you have to resolve your border issues?[49]

India's anxieties have quickly translated into North Atlantic anxieties because of its heft as a regional hegemon, its economic might, and its geostrategic significance. In the aftermath of the devastating August 2010 flood in neighboring Pakistan, media reports emphasized the prospect for refugees to push into neighboring Iran, Afghanistan, and India.

The South Pacific offers a second regional example. A significant proportion of the membership of the Association of Small Island States (AOSIS) is located here. AOSIS is not new. It was established in 1990 and participated

48. Architesh Panda, "Climate Refugees: Implications for India," *Economic & Political Weekly* 45: 20 (May 15, 2010), 76–79.
49. Tom Gjelten, "Pentagon, CIA Eye New Threat: Climate Change," National Public Radio, *All Things Considered*, December 14, 2009, available at www.npr.org.

in the 1992 Rio Summit, which recognized Small Island Developing States (SIDS) as a diplomatic entity. Yet, while island states almost always exhibit an asymmetrical interdependence with mainland economies, the scope of these states' precarious status became fully evident only in the '00s. Perhaps the starkest of environmental refugee images is that of an inundated island, its population forced into boats.

In 2001, it was reported that in response to that specter New Zealand had decided to extend immigration quotas to the island of Tuvalu. The report persisted throughout the decade. At the time, the Tuvalu government officially cited the step as evidence of a generous spirit typical of Pacific islands. This was a clear dig at Australia's Conservative government and Prime Minister John Howard. Others cast it as the Kiwi Labour government's way of embarrassing Howard. It all became moot, however, when New Zealand's Ministry of Foreign Affairs and Trade released a statement clarifying that it had not extended quotas to Tuvalu's citizens because of climate change but as part of an ongoing program of extending quotas to citizens of Pacific Access Category countries (for example, Fiji, Samoa, Tonga, Kiribati, and Tuvalu) to live and work in New Zealand.[50]

Nonetheless, the image of climate refugees emerging from inundated Pacific (and Caribbean) islands in the Pacific (and Caribbean) became increasingly salient in the '00s. By the end of the decade, beyond scholarly analyses, artists were dramatizing the peril. Films have begun to take up the issue of climate migrants directly. Michael Nash's *Climate Refugees* was screened at the 2010 Sundance Film Festival.[51] The documentary has been screened at other film festivals but has not been released to the public. Its trailer features ominous music and stark images of flooded islands. Then U.S. Speaker of the House Nancy Pelosi and Senator John Kerry are interviewed, with both arguing that "climate refugees" are a national security concern. Jennifer Redfearn and Tim Metzger's *Sun Come Up*, which focuses on the Carteret Islands in the Pacific, appeared as a trailer at various festivals in 2009 and was featured at Columbia University during New York City's Climate Week in 2009.[52] Outside the context of inundated islands, Abderrahmane Sissako's 2006 film *Bamako* depicts the fictionalized trial of the World Bank and the IMF, held in a courtyard in the Malian capital. A central testimony is that of a young man who had traveled to Morocco via Niger and Algeria only to be expelled by Moroccan authorities back to

50. Text available at www.mfat.govt.nz/Foreign-Relations/Pacific/NZ-Tuvalu-immigration.php.
51. The film's website is www.climaterefugees.com.
52. The film's website is www.suncomeup.com.

Algeria. The testimony was accompanied by a dramatic reenactment of people walking across stark desert landscapes.

A third emergence in the last decade—germane to the African context—was in Southern Africa. For example, in 2000, 2001, and 2007 devastating floods bedeviled the Zambezi and Limpopo river basins in Mozambique.[53] The cruel climatic seesaw between drought and floods in the country is problematic, to say the least, and couples with coastal soil erosion and rising sea levels in the delta regions.[54] Given the impact of the ENSO on the Indian Ocean, there is reason to assume that the region will remain vulnerable. Not only have people lost their housing, they have also lost their abilities to grow food. To date, there have been no major international migrations resulting from the Zambezi flooding.[55] Nonetheless, anti-immigrant sentiment has grown in recent years in South Africa, as has a control ethos devoted to tightened border control that, some argue, is borrowed from the North Atlantic. There is no reason to expect that future sentiment will not incorporate environmental security as an ancillary discourse.[56]

THE NORTH ATLANTIC SECURITY COMMUNITY

Other regional dynamics more pertinent to this book are evident in the North Atlantic. As demonstrated in chapter 4, transit states began to amp up the idea of climate refugees as a threat in the late '00s. In the Mediterranean, for example, transit states in the region have found CIM to be a useful new alarm bell to sound. Long adept at playing powerful cards—threats of Islamism, economic collapse, political instability—CIM may be an emergent trump card that builds on the already powerful immigration card. The point is that, simultaneously, CIM has been stirred into broader security imperatives designed to thwart irregular migration and cast as a new, deeper threat. It is an effective mobilizer. From the perspective of North Atlantic security officials, it has become a convenient hook on which to hang security rationales.

53. Alex de Sherbinin, Koko Warner, and Charles Ehrhart, "Casualties of Climate Change," *Scientific American* 304:1, January 2011, 64–71.

54. Marc Stal, "Mozambique," in *EACH-FOR: Environmental Change and Forced Migration Scenarios D.3.4. Synthesis Report*, ed. Andras Vag (Brussels, Belgium: European Commission, 2009), 40–41.

55. Ibid.

56. Megan Lindow and Alex Perry, "Anti-Immigrant Terror in South Africa," *Time*, May 20, 2008. See also Loren Landau and Aurelia Kazadi Wa Kabwe-Segatti, *Human Development Impacts of Migration: South Africa Case Study* (United Nations Development Programme Research Paper 2009/5).

Karl Deutsch and his colleagues conceived of the North Atlantic as a "security community" in the aftermath of World War II.[57] And there is significant sentiment among strategic policy makers that CIM poses a security threat to this community. At the same time, there are distinct differences between Europe and the United States—and, indeed, between Europe's southern and northern tiers—in their treatment of CIM. This discussion proceeds by treating the two continents separately before turning to the mutual response by NATO. As in India, it is clear that CIM was not initially treated as a threat per se; it was only as the '00s wore on that it became more prominent.

Europe

In Europe, a securitized response toward mixed flows of irregular immigration was evident in the early '00s with the Integrated System for the Surveillance of the Strait (SIVE) and Operation Ulysses. SIVE, based in Algeciras, Spain, combines radar, satellite, and motion detection systems. Begun by the Aznar government in 1999, it was expanded by 2003 to be a "fence in the water" between Algeria, Morocco, and Spain (including the Canary Islands).[58] It continues a rather robust operation, despite criticisms from NGOs and mounting evidence that smugglers are adept at bypassing the system. Perversely, such systems often reward successful smugglers.[59]

For its part, Operation Ulysses began in 2003 as an outgrowth of the cooperation between the Aznar, Blair, and Berlusconi governments in the aftermath of the 2002 Seville Summit and their participation in the "Coalition of the Willing" against Iraq. At the summit, immigration was cast as a

57. Karl Deutsch, *Political Community and the North Atlantic Area: International Organization in the Light of Historical Experience* (Princeton, NJ: Princeton University Press, 1957). I treat similar concerns in Gregory White, "Free Trade as a Strategic Instrument in the War on Terror? The 2004 U.S.-Moroccan Free Trade Agreement," *Middle East Journal* 59: 4 (Fall 2005), 597–616.

58. Jørgen Carling, "Migration Control and Migrant Fatalities at the Spanish-African Border," *International Migration Review* 41: 2 (2007), 316–343.

59. See also David Kyle and Rey Koslowski, eds., *Global Human Smuggling: Comparative Perspectives* (Baltimore, MD: Johns Hopkins University Press, 2001); Rey Koslowski, "The Mobility Money Can Buy: Human Smuggling and Border Control in the European Union," in *The Wall around the West: State Borders and Immigration Controls in North America and Europe*, eds. Peter Andreas and Timothy Snyder (Lanham, MD: Rowman & Littlefield, 2000), 203–218; and Wayne Cornelius, "Death at the Border: Efficacy and Unintended Consequences of U.S. Immigration Control Policy," *Population and Development Review* 27: 4 (2001), 661–685.

wholesale security concern. The UK and Spain spearheaded the effort to place patrol boats in the Mediterranean, but France, Portugal, and Italy also joined the endeavor. Greece, Norway, Germany, Poland, and Austria participated as observers. The effort was short-lived because of the onset of the Iraq War in March 2003, but it was resuscitated in 2006, with European boats patrolling the Atlantic waters around the Canary Islands. Other patrol operations have included France's Operation Amarante in 2002–2004.[60] NATO's Operation Active Endeavor is discussed below.

In October 2005, the European Agency for the Management of Operational Cooperation at the External Borders of the Member States (Frontex) came into operation. Tasked with integrated border management and the implementation of the Schengen Acquis, Frontex quickly became a hefty player on the scene. It inherited responsibility for coordinating joint operations, such as Nautilus and EPN-Hermes in 2008. And it is responsible for reporting the situation at the EU's external borders. One of the striking aspects of Frontex is that its militarized and robust presence exhibits a striking degree of supranational and multilateral cooperation. According to its chair, Robert Strondl, Frontex has sought to broker relationships in a "network approach" with Europol, UNHCR, IOM, the European Maritime Safety Agency (EMSA), Interpol, International Centre for Migration Policy Development (ICMPD), and the European Police College (CEPOL).[61] Frontex does not seem to amplify the security threat from CIM; it is more about implementing policy and leaves the overt security discourse to think tanks and policy makers.

At the diplomatic level, in 2001 the European Commission began a systematic approach to supporting third-country transit states in their efforts to mitigate migration pressures. By 2004, the EU had put in place Aeneas, a rubric for providing financial and technical assistance to nonmember countries. (Aeneas was the hero of Virgil's *Aeneid* and a key character in Homer's *Iliad*, skilled in fighting and devoted to duty.) At the heart of Aeneas were 107 projects designed to deepen international networking on immigration control, protection frameworks, and the interdiction of illegal migration. Two examples illustrate the nature of the projects. In 2005, a €1.9 million project (about US$2.6 million at the time) was funded to support cooperation between Libya and Niger in border control, "with special reference to irregular migratory flows from sub-Saharan Africa transiting

60. Derek Lutterbeck, "Policing Migration in the Mediterranean," *Mediterranean Politics* 11: 1 (March 2006), 59–82.

61. Robert Strondl, *Frontex: General Report for 2008* (Warsaw, Poland: Frontex, 2008), available at www.frontex.europa.eu.

the two countries to reach the coasts of Southern Italy and then other European countries."[62] The implementing agency was the Department of Public Security in Italy's Ministry of Interior. No evaluation of the project appears to have been conducted. In 2006, the same department was central to a second project for €1.2 million (about US\$1.4 million at the time) to work with Algeria and Libya against illegal migration. The experience of countries like Libya is instructive. Prior to January 2011 and Libya's collapse, Qaddafi had been welcomed back into the international security fold after assiduous British diplomacy led to his renunciation of WMD in 2003. In addition to lucrative oil contracts, Italian and EU security officials had nurtured Libya's participation in migration interdiction efforts.[63]

As the decade moved on, calls for treating climate change as a security threat increased, with climate-induced migration at their very core. For example, the German Advisory Council on Global Change (known by its German acronym, WBGU) released a report in 2007 stating that climate change was "jeopardizing national and international security to a new degree."[64] It singled out the "conflict constellation" of "environmentally induced migration" that will emerge from climate change. To its credit, the WBGU notes that most migration occurs within countries and that much of it tends to be south–south. Yet it also argues, "Europe and North America must also expect substantially increased migratory pressure from regions most at risk from climate change." In some ways, perhaps, the report is rather contradictory. It repeatedly uses terms like "security," "threat," "risk," "conflict," and "dangerous," yet it eschews a draconian security response. It calls, instead, for efforts to establish a "cross-sectoral multilateral Convention aiming at the issue of environmental migrants." Such eschewal and the call for a multilateral, cooperative response is important. Still, in the context of this analysis, the overall tone of the WBGU document is noteworthy.

A second alarmed call came from a 2008 joint report, "Climate Change and International Security," prepared by EU High Representative for the Common and Foreign Security Policy Javier Solana and the European Commission.[65] Solana is a prominent figure in Europe, having served in several

62. European Commission, *Aeneas Programme: Overview of Projects Funded 2004–2006* (Brussels, Belgium: Europe Aid Programme of the European Commission, 2008).

63. Derek Lutterbeck, "Migrants, Weapons and Oil: Europe and Libya After the Sanctions," *Journal of North African Studies* 14: 2 (2009), 169–184.

64. German Advisory Council on Global Change, *World in Transition: Climate Change as a Security Risk* (Berlin, Germany: 2007).

65. Javier Solana and European Commission, "Climate Change and International Security: Paper from the High Representative and the European Commission to the European Council" (Brussels, Belgium: Council of the European Union, 2008).

high-profile roles, including Secretary General of NATO. Invoking climate change as a "threat multiplier," the report takes the blunt tact of derogating humanitarian matters as less important than security concerns. It argues, "It is important to recognize that the risks are not just of a humanitarian nature; they also include political and security risks that directly affect European interests."[66] It identifies climate change as an "amplifier" to "poor health conditions, unemployment or social exclusion" and states vaguely that there will be "millions of environmental migrants" by 2020. Compared to the U.S. response (discussed below), the Solana document, like its WBGU counterpart, is more adept at invoking multilateral cooperation. Nevertheless, the militarized backdrop, including the border control efforts, remains the actual, on-the-ground manifestation of policy advocacy. Additionally, the calls for multilateral cooperation are preoccupied with "the security risks related to climate change in the multilateral arena; in particular within the UN Security Council, the G8, and the UN specialised bodies" and the need to enhance "cooperation on the detection and monitoring of the security threats related to climate change."

United States

CIM was not an issue for most of the '00s because of the Bush-Cheney administration's aforementioned positions on the environment and climate change. Without climate change itself as a concern, it would have been hard for administration officials to argue for a securitization of CIM. One might argue that the aggressive policy of securing the homeland in the aftermath of September 11 and being "tough" on illegal immigration would guard against immigrants motivated to move by environmental factors, too. Still, neither the White House nor the intelligence and military bureaucracies seemed intent on environmental security, never mind CIM.

This changed with the 2006 election of Democrats to the House of Representatives and the Senate. In the spring of 2007, the House of Representatives included in its 2008 intelligence authorization a directive for the National Intelligence Council (NIC) to produce a National Intelligence Estimate (NIE) on climate change.[67] An NIE is the intelligence community's assessment of a specific national security issue. House Republicans, led by Peter Hoekstra (R-MI), ranking minority leader and former chair of the

---

66. Ibid.
67. Walter Pincus, "Intelligence Chief Backs Climate Study: McConnell Calls Security Review 'Appropriate'; some GOP Leaders Oppose Idea," *Washington Post*, May 12, 2007.

House Permanent Committee on Intelligence, were irate that intelligence resources would be devoted to such an effort, arguing that it would take attention away from more pressing security concerns. They also protested a congressional committee mandating action by the NIC. Nevertheless, National Intelligence Director Mike McConnell, appointed by President Bush in January 2007, defended the directive as an appropriate step for the NIC. Throughout 2007 and 2008, the presidential campaign loomed large in the U.S. Senate, and climate security remained very much in the mix in Washington. Ultimately, the Lieberman-Warner Climate Security Act was withdrawn in June 2008 because it did not have sufficient votes to reach cloture. Climate security, in this context, was cast as more about economic considerations and energy self-reliance than climate change per se, but it was part of the picture.

NIC Chairman Thomas Fingar reported the NIE findings to Congress in June 2008 during the deliberation of the climate change bill.[68] The NIE concluded that climate change would challenge U.S. national security, especially in sub-Saharan Africa, the newest military command for the Pentagon. Secretary of Defense Robert Gates had established African Command (AFRICOM) in February 2007. The intelligence community, Fingar reported, had concluded that humanitarian disasters, economic migration, and food and water shortages would be caused by climate change and would "tax U.S. military transportation and support force structures, resulting in a strained readiness posture."[69] Fingar said Africa was most vulnerable "because of multiple environmental, economic, political and social stresses." While no country would avoid climate change, the report said, it identified "most of the struggling and poor states that will suffer adverse impacts to their potential and economic security" as being in the Middle East, central and southeast Asia, and sub-Saharan Africa. The United States must "plan for growing immigration pressures," the report said, too, in part because almost a fourth of the countries with the greatest percentage of low-level coastal zones are in the Caribbean. The report noted that many U.S. military installations near the coast would be at "increasing risk of damage" from floods in coming years.

Similarly, and with respect to CIM, Director of National Intelligence Dennis Blair, who succeeded Mike McConnell, reported in his annual threat assessment in February 2009 that the intelligence community expects that

68. Permanent Select Committee on Intelligence and House Select Committee on Energy Independence and Global Warming, "Testimony by Thomas Fingar, Deputy Director for National Intelligence, on the National Security Implications of Global Climate Change to 2030."
69. Ibid.

"economic migrants will perceive additional reasons to migrate because of harsh climates, both within nations and from disadvantaged to richer countries."[70] The use of "economic migrants" as a category is politically intriguing, of course. Whether it is a careless interchange or a specific one is impossible to know. One suspects that it is intentional, given that many eyes redact such documents. Casting CIM as ultimately *economic*, with "harsh climates" being "perceived" as a "reason to migrate," is a deft way of emphasizing a migrant's volition.

Additionally, an important entry in the discourse was *Global Trends 2025: A Transformed World*, the fourth installment of the NIC's effort to assess future scenarios.[71] It emerged in November 2008, as the Bush-Cheney administration was coming to an end. It is especially significant because it names "climate migrants"—via the 2006 Stern Report—as an explicit security concern. *Global Trends 2025* offers a fascinating, provocative set of arguments. First, it argues that climate change and CIM are not security concerns so much as they *will be invoked* as security concerns, setting in motion an unfortunate dynamic. It is an effective rhetorical device: "I'm not saying X, but there are those who do or will say it" is surely a clever way of saying X. This further prompts anxieties about the CIM security dilemma, wherein one country's effort to securitize CIM leads another to do the same:

> Over the next 20 years, worries about climate change effects may be more significant than any physical changes linked to climate change. Perceptions of a rapidly changing environment may cause nations to take unilateral actions to secure resources, territory, and other interests. Willingness to engage in greater multilateral cooperation will depend on a number of factors, such as the behavior of other countries, the economic context, or the importance of the interests to be defended or won.[72]

Despite the rhetorical flourish, this is an intriguing meta-level of awareness on the part of the NIC. It is also a provocative point of discussion. With respect to CIM, as discussed in chapter 5, the importance of multilateral cooperation outside a militarized mind-set may be absolutely essential to avoid a CIM security dilemma.

Global Trends 2025 goes further, too, in pointing out the security implications of GHG mitigation and the potential for punctuated events:

70. Senate Select Committee on Intelligence, *Annual Threat Assessment of the Intelligence Community*, Washington, DC: U.S. Senate. 2009.

71. National Intelligence Council, *Global Trends 2025: A Transformed World* (Washington, DC: U.S. Government Printing Office, 2008).

72. Ibid., 53–54.

Many scientists worry that recent assessments underestimate the impact of climate change and misjudge the likely time when effects will be felt. Scientists currently have limited capability to predict the likelihood or magnitude of extreme climate shifts but believe—based on historic precedents—that it will not occur gradually or smoothly. Drastic cutbacks in allowable CO_2 emissions probably would disadvantage the rapidly emerging economies that are still low on the efficiency curve, but large-scale users in the developed world—such as the US—also would be shaken and the global economy could be plunged into a recession or worse.[73]

Assessing the costs of mitigation is an important task for the intelligence community. One hopes that the anxieties expressed do not abet efforts to avoid cutbacks.

The full impact of the change in administration in January 2009 still remains to be seen in terms of climate security; the effect on climate and energy policy of the 2010 Deepwater Horizon tragedy in the Gulf of Mexico is uncertain. Nonetheless, three developments in the first half-term of the Obama-Biden administration merit mention. First, in October 2009, the Central Intelligence Agency established a new Center for the Study of Climate Change. Second, in 2010, for the first time, climate change was among the security threats identified by Pentagon planners in the Quadrennial Defense Review. Third, the administration facilitated climate scientists' use of defense and intelligence data. This practice of allowing scientists to use state-of-the-art satellite and other surveillance data—for example, to gauge the loss of Arctic ice—began in the '90s under the Clinton-Gore administration and was stopped during the Bush-Cheney administration. The navy needs such data for strategic calculations about the Arctic Ocean, but it is also a treasure trove for scientists. Per the usual debate, Republican members of Congress have criticized the practice as a waste of resources. Senator John Barrosso (R-WY) argued, "We shouldn't be spying on sea lions."[74]

NATO

In geostrategic and military calculations, NATO's efforts often reveal commonality in EU and U.S. policy. In recent years there has been a deepening of trans-Atlantic cooperation on irregular immigration. U.S. military

73. Ibid.
74. William J. Broad, "C.I.A. Is Sharing Data with Climate Scientists," *New York Times*, January 4, 2010.

efforts in North America are unilateral, of course, but in the Mediterranean and Eastern Europe they are multilateral and conducted under the aegis of NATO. While the military and patrol operations conducted in the early '00s were typically European-based, by the middle of the decade they were NATO-based. Still, most of the operations do not appear to involve too much by way of U.S. assets. Coordination with Frontex has been a constant source of new deliberations since 2005.

Operation Active Endeavor, for example, came into existence after September 11 and was fully operational in April 2003. Based at the Joint Forces Command in Naples, it has been a presence in the Mediterranean basin from the Strait of Gibraltar through the narrow body of water separating Sicily from Tunisia's Cap Bon. Beginning with the June 2004 Istanbul Summit, it was expanded all the way to the eastern Mediterranean. Its focus is on "antiterrorist" activities. Yet in its own description, NATO trumpeted Operation Endeavor's work vis-à-vis irregular immigration. "NATO ships and helicopters have also intervened on several occasions to rescue civilians on stricken oil rigs and sinking ships. This includes . . . winching women and children off a sinking ship carrying some 250 refugees in January 2002 and helping to repair the damaged hull."[75] The use of the word *refugee* is striking; again, whether it is intentional or inadvertent is impossible to know.

The bulk of the contribution to Operation Active Endeavor comes from Greece, Italy, Spain, and Turkey. Germany, Denmark, and Norway have contributed as well. Intriguingly, non-NATO countries are participating: Russia and Ukraine have contributed assets since 2007, and Israel, Morocco, and Georgia have signed letters of intent. In October 2009, Morocco signed in Naples a "Tactical Memo of Understanding" for a Moroccan contribution.[76]

Throughout the '00s, the notion of environmental security became more and more evident in NATO's strategic planning. Most of the attention was on mining, pollution mitigation, pesticide reduction, and natural disasters; the CIM threat was only indirect. As the decade drew to a close, however, there was evidence that NATO was emphasizing CIM, too. At the March 2008 NATO Security Science Forum in Brussels, it received close attention.[77] Using the *NATO Review* as an indicator, by 2008 and 2009 the neglect of CIM was a thing of the past. For example, in the 2009 *NATO Review*, Achim Steiner echoed Myers-style estimates:

75. See www.nato.int/cps/en/natolive/topics_7932.htm#contributing.

76. Press release available at www.jfcnaples.nato.int/organization/CC_MAR_Naples/PressReleases/CC-MAR/pressreleases09/NR_20_09.html.

77. See www.nato.int/docu/comm/2008/0803-science/0803-science.htm.

Forecasts on the number of persons that might have to migrate due to climate change and environmental degradation by 2050 vary between 50 million and 350 million. Environmental change will impact migration in three ways. First, global warming will decrease agriculture potential and undermine water availability. Second, the intensification of natural hazards such as flood, storm or drought, will affect more and more people with low adaptive capacity and generate uncontrolled large-scale human displacement. Third, densely-populated and low-lying coastal areas will be permanently flooded or damaged leading to relocation without return, recovery, and reintegration possible. A one meter sea level rise would result in the entire disappearance of the Maldives for example. Both Kiribati and the Maldives have ongoing resettlement plans.[78]

NATO and Frontex's roles in the securitization of CIM continue to evolve and deepen. With the intervention in Libya in March 2011, the European Council requested at its March 24-25 meeting that Frontex step up its presence in the Mediterranean.

Given present trends, North Atlantic military attention to climate security is likely to accelerate in the years to come. Climate change challenges military planners in three areas. First, there are the operational challenges posed to infrastructure by a changing world. As sea levels rise due to thermal expansion and melting ice packs, port access might change, and island bases will be challenged. Second, the geopolitical dimension is real, as resource wars become a greater concern.[79] Third, the strategic challenge posed by relocated peoples is significant. Military strategists are now examining these challenges in ways that were inconceivable back in the late '90s.

Finally, within the context of NATO, an important dimension to the expansion of security efforts with respect to CIM is the incorporation of Maghrebi, Sahelian, and sub-Saharan countries in the security community. In the early '00s, the United States initiated the Pan-Sahel Initiative (PSI) and the Trans-Sahara Counterterrorism Initiative (TSCTI). The PSI began in 2002 and was succeeded by the TSCTI in 2005.[80] They led to military exercises such as 2005's Operation Flintlock, a joint endeavor with participation by Algeria, Chad, Burkina Faso, Libya, Mali, Mauritania, Morocco, Niger, Nigeria, Senegal, and Tunisia. Part of the impetus for the TSCTI was

78. Achim Steiner, "Environment as a Peace Policy," *NATO Review: How Does NATO Need to Change (Parts 1 and 2)?*, available at www.nato.int/docu/review.

79. Michael Klare and Daniel Volman, "The African 'Oil Rush' and US National Security," *Third World Quarterly* 27: 4 (2006), 609–628.

80. David Gutelius, "Islam in Northern Mali and the War on Terror," *Journal of Contemporary African Studies* 25: 1 (2007), 59–76.

surely counterterrorism, with additional attention to securing oil assets in the region.[81] The implications of this extension of the North Atlantic security community's purview remain to be discerned. The problem for now is that governments seen to be collaborating with the U.S. and Europe have had their legitimacy undermined. The countries' "coercive architectures" are enhanced, thereby broadening the support for oppositional forces and political activity easily framed as terrorist by the North Atlantic.[82]

A conceptualization of the Sahara and the Sahel as ungoverned spaces filled with terrorists and drug traffickers has been at the heart of this effort. Surely there are unsavory, criminal elements in the region, yet wholesale military operations, without an attention to developmental and environmental policies, are misguided. Such assertions are part of a long-held practice in state formation and developmentalism, as the state seeks to extend its purview into new territory deemed as open frontier. Certainly the spaces are not empty. They are only without a securitized presence.

UN SECURITY COUNCIL, THE WORLD BANK, AND NGOs

Before concluding, a consideration of the UN Security Council and the role that key NGOs and think tanks are playing is in order. The Security Council took up the security dimension of climate change in a landmark session in April 2007. Not surprisingly, it examined the issue from the perspective of "collective security." After all, that is the Security Council's ambit under Chapter VII of the UN Charter. Interestingly, the council's consideration took on the language of warfare, especially in testimony offered by southern states. Representatives from Tuvalu equated greenhouse gases to WMD, with chimney stacks and exhaust pipes as weapons. At the same time, the Security Council debate tended to eschew environmental-conflict discourse; it resided more broadly in the less draconian environmental-security realm.[83]

In June 2010, the World Bank, too, joined in on the growing preoccupation with CIM by hosting the First Annual Workshop on Climate-Induced

---

81. Klare and Volman, "The African 'Oil Rush' and US National Security."

82. Cédric Jourde, "Constructing Representations of the 'Global War on Terror' in the Islamic Republic of Mauritania," *Journal of Contemporary African Studies* 25: 1 (January 2007), 77–100.

83. Detraz and Betsill, "Climate Change and Environmental Security: For Whom the Discourse Shifts," 311.

Migration and Displacement in the Middle East and North Africa. Held at the Center for Mediterranean Integration in Marseilles, France, and cosponsored with the French Development Agency (AFD), the conference featured regional specialists offering analyses concerning the expected impact of climate change on demography and migration. The newness of the attention was trumpeted repeatedly.

Other actors are relevant in considering the North Atlantic response, especially NGOs. There are two kinds of NGOs in this context. First, aid organizations and/or transborder advocacy networks concerned about the humanitarian and ethical implications of CIM are an important presence.[84] These include CARE, Christian AID, Human Rights Watch, and the Norwegian Refugee Council. But a second kind of player is also pertinent: NGOs or think tanks that involve former military officials and policy makers, including the Center for a New American Security (CNAS), the Center for Naval Analyses (CNA), the American Security Project (ASP), CENTRA Technology, Inc., the Royal United Services Institute (RUSI), and other NGOs and for-profit and nonprofit firms in Washington and European capitals. Their recent, sustained attention to environmental security and CIM is noteworthy.

A landmark, and extreme, view of the impact of climate change on national security appeared in a widely cited 2003 report prepared by Schwartz and Randall for the Department of Defense's Office of Net Assessment.[85] The primary national security challenge in a world of climate change, they argued, was "border management" against desperate aspirants. It argued:

> With diverse growing climates, wealth, technology, and abundant resources, the United States could likely survive shortened growing cycles and harsh weather conditions without catastrophic losses. Borders will be strengthened around the country to hold back unwanted starving immigrants from the Caribbean islands (an especially severe problem), Mexico, and South America.[86]

"Hold back unwanted starving immigrants." The report was widely cited at the time. Whether or not it enjoyed widespread acceptance is hard to know.

---

84. Margaret Keck and Kathryn Sikkink, *Activists beyond Borders: Advocacy Networks in International Politics* (Ithaca, NY: Cornell University Press, 1998).

85. Peter Schwartz and Doug Randall, *An Abrupt Climate Change Scenario and Its Implications for United States National Security* (New York: Global Business Network, 2003), 22.

86. Ibid., 18.

Nonetheless, such reports often shape official outlooks and can be revealing in terms of discursive changes and policy evolution.

Beginning in 2007 there was a sharp increase in game simulations, modeling, and scenario-building exercises coinciding with the advent of the Democratic-controlled Congress. A seminal example of this effort was a collective product of the Center for Strategic and International Studies (CSIS) and the Center for a New American Security (CNAS). Entitled *The Age of Consequences: The Foreign Policy and National Security Implications of Global Climate Change*, the report brought together an impressive array of former officials, including former CIA Director James Woolsey, former Chief of Staff (to President Clinton) John Podesta, former Deputy Assistant of Defense Kurt Campbell, and former National Security Adviser to the Vice President (Gore) Leon Furth.[87] The title evokes Churchill's warning against complacency in the late '30s. It is a thorough analysis whose perspective is sober yet very alarmed. Climate change, it argues, inescapably poses a threat to national security, one "as great or greater" than Iraq, Afghanistan, energy security, violent extremism, and natural or man-made pathogens. Its argument with respect to CIM merits extensive quoting:

> Perhaps the most worrisome problems associated with rising temperatures and sea levels are from large-scale migrations of people—both inside nations and across existing national borders. [In all foreseeable] scenarios it was projected that rising sea levels in Central America, South Asia, and Southeast Asia and the associated disappearance of low lying coastal lands could conceivably lead to massive migrations—potentially involving hundreds of millions. These dramatic movements of people and the possible disruptions involved could easily trigger major security concerns and spike regional tensions. In some scenarios, the number of people forced to move in the coming decades could dwarf previous historical migrations. The more severe scenarios suggest the prospect of perhaps billions of people over the medium or longer term being forced to relocate. The possibility of such a significant portion of humanity on the move, forced to relocate, poses an enormous challenge even if played out over the course of decades.[88]

Of the CSIS/CNAS document's enumeration of the 10 consequential implications of climate change, CIM is second after increased north-south tensions. In the end, the report's thoroughness and analysis are striking.

87. Kurt Campbell et al., *The Age of Consequences: The Foreign Policy and National Security Implications of Global Climate Change* (Washington, DC: Center for a New American Security and Center for Strategic and International Studies, 2007).
88. Ibid., 9.

Fascinating, too, is its epigraph, which quotes Machiavelli approvingly. The authors do not quote the Machiavelli of *The Discourses*, with its vigorous republicanism, but the cynical, manipulative Machiavelli of *The Prince*. History has forgotten that with *The Prince* Machiavelli may have sought to prompt the downfall of his loathsome patron, Medici.[89] Rousseau also pointed out there is a sharp difference between the two Machiavellis.[90] The authors of *The Age of Consequences* chose the cynical Machiavelli—a passage from chapter 3 of *The Prince*:

> As the physicians say . . . in the beginning of [a] malady it is easy to cure but difficult to detect, but in the course of time, not having been either detected or treated in the beginning, it becomes easy to detect but difficult to cure. Thus it happens in affairs of state, for when the evils that arise have been foreseen (which it is only given to a wise man to see), they can be quickly redressed, but when, though not having been foreseen, they have been permitted to grow in a way that every one can see them, there is no longer a remedy.[91]

It is an unintended and intricate irony for security officials to quote approvingly a chapter from *The Prince* that focuses on the foreign policies of the Greek and Roman Empires. Machiavelli knew history well. Empires ultimately fall.

This caution against waiting until a security crisis fully emerges appears to be at the heart of security officials' calculations. To its credit, the CSIS/CNAS document does not go as far as Schwartz and Randall's 2003 document, with the latter's alarmism and cynical advocacy of harsh CIM interdiction. Nevertheless, it eschews a rigorous multilateralism or other solutions.

89. Mary Dietz, "Trapping the Prince: Machiavelli and the Politics of Deception," *American Political Science Review* 80 (1986), 777–799.

90. Jean-Jacques Rousseau, *On the Social Contract*, trans. Judith R. Masters (New York: St. Martin's, 1978). Rousseau writes: "Machiavelli was an honorable man and a good citizen; but being attached to the Medici household, he was forced, during the oppression of his homeland, to disguise his love of freedom. The choice of his execrable hero is in itself enough to make manifest his hidden intention; and the contrast between the maxims of his book *The Prince* and those of his *Discourses on Titus Livy* and of his *History of Florence* shows that this profound political theorist has had only superficial or corrupt readers until now. The court of Rome has severely forbidden his book. I can well believe it; it is the court that he most clearly depicts" (88).

91. Campbell, *The Age of Consequences: The Foreign Policy and National Security Implications of Global Climate Change*, 5.

States make choices regarding policy. Not always as they please, of course, as options can be sharply constrained depending on historical legacy, bureaucratic politics within the state, domestic and societal pressures, and geopolitical position. Still, states make crucial policy decisions, and in the realm of CIM, choices are being and will be made.

The fundamental argument here is that the environment has been securitized and, moreover, that the securitization of CIM dovetails with anti-immigration policies and climate change security. Building fences against irregular migration is politically successful. Ironically, too, a public that is skeptical about climate change or unconvinced that it poses a threat to their lives finds a precautionary principle at work in securitizing CIM. "We're not sure that climate change is a threat," they might reason, "but just in case it is, let's stop immigrants who are moving because of climate change." Finally, extending the purview of the security state into the apparatuses of transit states may seem prudent, too. Table 3.1 demonstrates the matrix at issue.

At bottom, however, the long-term costs and/or the ultimate inefficacy of the steps is not well considered. Mitigation of GHGs may indeed be an expensive proposition, unless political leaders craft it as an opportunity to dynamize the world economy with innovative technologies.

As the next chapter turns to transit states, the CIM security dilemma remains pertinent. If one country seeks to protect itself against CIM, its actions will run against the interests of another. As Arizona guards itself, New Mexico feels vulnerable. A U.S. effort deepens Mexico's anxieties

Table 3.1 COST-BENEFIT OF POLICY STEPS

	Perceived cost	Actual cost	Political payoff	Long-term efficacy
Mitigation of GHGs	High	Medium*	Low	High
Adaptation of affected populations	High	Low	Low	High
Building fences	Low	High	High	Low
Enhancing capacity of transit states	Low	High	High	Low

about its ability to absorb CIM from elsewhere. If Mexico securitizes CIM, what impact does that have on Belize or Guatemala? Will transit migrants settle in Belize instead? Will Guatemala build walls on its borders with El Salvador and Honduras? If the Spanish coast is perceived (rightly or wrongly) as less passable, what are the implications for Italian security? (Both countries are members of the EU, of course, so what are the implications for Frontex and other supranational cooperation efforts?) How will European efforts affect Morocco and Tunisia and Libya? Will Maghrebi countries harden their southern borders? If Morocco continues its toughening of policy, what does that mean for neighboring Algeria and its immigrants? What will this mean for Sahelian countries to the south?

To return to Machiavelli's sycophantic advice to the authoritarian Medici, quoted so approvingly by a group of prominent U.S. policy intellectuals: Is doing something before others recognize the problem the solution? Is that really the best form of governance?

CHAPTER 4

Transit States and the Thickening of Borders

The securitization of migration—the complicated effort to control the mixed flows of migrants seeking to enter North Atlantic countries—has deepened in recent decades. Climate change and security concerns have now fully merged in an increased elaboration of anxieties concerning climate-induced migration. As detailed in chapter 3, this has not only changed the discourse of advanced industrialized countries vis-à-vis mixed migration flows but has also involved actual efforts to build fences, patrols, and detention centers to thwart migrants.

The endeavor to inject security imperatives into migration control has also involved much more, namely a preemptive extension of interdiction efforts beyond North Atlantic borders. U.S. and EU officials have devoted resources to encouraging and pressing neighboring countries to thwart migrants' passage. The resulting "border thickening" extends North Atlantic sovereignty into other regions. The U.S. Customs and Border Protection (CBP) agency, which occupies fully one-fourth of Department of Homeland Security (DHS) personnel, has increasingly conflated immigration, terrorism, and drug control. In testimony before the House Select Committee on Homeland Security in 2004, CBP director Robert Bonner said homeland security includes border protection programs that "have been put in place at—and beyond—our borders." He added, "The very existence of CBP makes us vastly better able to protect our nation from all external threats, whether illegal migrants and illegal drugs, terrorists, terrorist weapons, including weapons of mass destruction."[1]

---

1. Quoted in Tom Barry, *Pushing Our Borders Out: Washington's Expansive Concept of Sovereignty and Security* (Silver City, NM: International Relations Center, 2005).

The result has been a deterritorialization of migration controls ⟨ actors situated strategically within a migration system not yet fully p. the North Atlantic grouping. Countries such as Mexico, Morocco, Tunisia, Libya, and Turkey have, in different ways, been enlisted in border control efforts. As such, they have taken on the role of transit states. Being key pivot points, they are essential to destination countries' efforts to enhance control. Their repression of transmigration supplements North Atlantic efforts to proscribe immigration.

Yet such roles need to be unpacked, too, because it would not do to consider transit states as unitary, cohesive actors. Interestingly, transit states are embracing the role of mixed-flow interdiction. From an old-school dependency perspective, one might have argued that "core" countries coerce "third world" countries to pursue policies against their interests.[2] To the contrary, as argued here, transit states willingly help to thwart mixed flows of migration. In this regard, countries play complicated roles in "double-edged" diplomatic games.[3] States seek to negotiate and gain diplomatically while fully aware of their counterparts' domestic interests. And, in the case of irregular migration, transit states can actually emphasize and capitalize on the asymmetrical interdependence. They can play immigration as a diplomatic card in negotiations over other thorny issues. With the emergence of CIM as a security concern, this diplomatic card has become even more potent. This evolution of transit states has transformed modes of governance as well as sovereignty dimensions in profound ways.

Issues concerning transit states have received some attention in the scholarly literature, but not a great deal. And their emergence in the context of CIM has been largely ignored. To illuminate these matters, this chapter explores the case of Morocco, a central player in Mediterranean and Maghrebi politics. Morocco plays a somewhat different role than its Mediterranean counterparts, Libya and Turkey, or its North American analog, Mexico. Obviously, each context has its own circumstances. Nonetheless, Morocco's role as a transit state is representative.

More to the point for our concerns here, as the mixed flows increasingly include climate-induced migrants—or are perceived to—transit states'

2. J. Samuel Valenzuela and Arturo Valenzuela, "Modernization and Dependency: Alternative Perspectives in the Study of Latin American Underdevelopment," *Comparative Politics* 10: 4 (1978), 535–557; and Immanuel Wallerstein, *The Capitalist World Economy* (London: Cambridge University Press, 1979).

3. Peter B. Evans, Harold K. Jacobson and Robert Putnam, eds., *Double-Edged Diplomacy: International Bargaining and Domestic Politics* (Berkeley: University of California Press, 1993).

willingness to play this security role has only grown. Transit states are invoking climate-induced migration to amplify crucial strategic positions and will likely continue to do so. Morocco hosted the September 2009 meeting of the International Union for the Scientific Study of Population (IUSSP). It was the first time the renowned body of demographers and population specialists had held its meeting in an Arab or African country. The location was never lost on participants. A giant portrait of King Mohammed VI hung behind the podium in the main auditorium and was, therefore, featured in the broadcast of speeches on monitors above the stage and in the news media. The king's welcoming speech—delivered by High Commissioner for Planning Ahmed Lahlimi Alami—could not have made the point concerning Morocco's willingness to work with outside players to thwart CIM more explicitly:

> Food security, desertification, farmland depletion and sea level rise are pressing issues for demographers to ponder . . . so as to address their severe implications, particularly the serious problem of migration movements, a phenomenon which is set to take a more critical turn in the future. Located at a point where North meets South, Morocco feels very much concerned about migration issues. To face up to the impact of migration movements, we have been working with our Euro-Mediterranean partners on the strategies that need to be developed to bring this phenomenon under control.[4]

In this and other public discourses, Morocco's king repeatedly emphasizes the emergent threat posed by climate change, linking it directly to migration on the African continent. At the Third Africa-EU Summit, held November 2010, in Tripoli, Libya, Mohammed VI expressed concern about "security of the countries in the Sahel-Saharan zone" and then immediately invoked the importance of addressing migration and climate change.[5]

This chapter seeks to illuminate transit states' willingness to play the role of an external buffer against real, and perhaps imagined, flows of CIM. One would search in vain to find an evidentiary smoking gun: no transit state official would admit on the record that CIM is cynically emphasized as a means of securing support from North Atlantic officials. There is no denying, however, that transit states are part of a broader international

---

4. King Mohammed VI, "Speech to the 2009 Meeting of the International Union for the Scientific Study of Population," Marrakech, Morocco, September 27, 2009.
5. King Mohammed VI, "Speech to the 3rd EU-Africa Summit hosted by Libya," Tripoli, Libya, November 29, 2010.

security framework that sees CIM as an emergent threat. The fundamental argument of the chapter is that the transit state's role in a migration system has implications for international governance. Specifically, the security apparatus of the international state system is strengthened as each transit state is further incorporated into, as Weber would have it, the "relations of authority" of advanced-industrialized countries. States do not act as discrete national units, with a thin-line border marking the edge of their legitimate use of violence. The participation of transit states in migration interdiction deepens the transnationalization of the security state. Supranational cooperation between respective countries' security officials concerning mixed flows of irregular immigration (including CIM) strengthens transnational security. This is somewhat analogous to the transnationalism evident in Keck and Sikkink's examination of transnational advocacy networks of NGOs or Haas's treatment of epistemic communities focused on environmental issues.[6] It is especially relevant to Slaughter's analysis of transnational government networks.[7] Obviously, cooperation between U.S. and Mexican security officials—or between Italian and Libyan ministries of interior, or NATO and Moroccan navies—has implications for the migrants and refugees seeking entrance and encountering the coercive measures. Yet it also affects transit states' domestic politics, as well as the character of nationhood and national security for citizens within North Atlantic countries. The chapter concludes with a discussion of evolving sovereignty norms and the ways in which the securitization of immigration and CIM affects governance.

TRANSIT STATE DEFINED

Despite the valuable research done on migration flows and transit migration, the role of transit states in this process has not been well examined. Research has conventionally been devoted to immigration flows as a bilateral concern between pull (destination) and push (sending) countries. Most attention has focused on efforts by destination countries to control immigration or on their experiences as host countries contending with

6. Peter Haas, *Saving the Mediterranean: The Politics of International Environmental Cooperation* (New York: Columbia University Press, 1990); and Keck and Sikkink, *Activists beyond Borders: Advocacy Networks in International Politics* (Ithaca, NY: Cornell University Press, 1998).

7. Ann-Marie Slaughter, "The Real New World Order," *Foreign Affairs* 183, September/October 1997.

issues of assimilation or integration.[8] Less common is work that treats sending countries and their efforts to encourage emigration.[9] Sociological and anthropological works have added important insights concerning networks.[10] Regardless of the approach, a migration system is conventionally seen as a two-country dyad. A national departs from one country and arrives at a destination. Turkish nationals work in Germany, Mexicans in the United States, and Tunisians in France.

Of course, migrants often traverse countries that are not their final destination, posing challenges and concerns for those in-between countries. "Transit migrants" are often underground, moving clandestinely.[11] If they work in the country of transit, it is most likely without documentation, and they are unevenly integrated or assimilated into society, if at all. Their ultimate destination is elsewhere.

By definition, then, transit states occupy a pivotal position within a broader migration system in which a given state experiences and seeks to respond to changing migration patterns. These responses are invariably complicated, sometimes contradictory, and sometimes muddled. As detailed below, they selectively encourage or discourage significant immigration and emigration. And for transit states, the conventional dichotomy between labor exporter (sending country, country of emigration) and labor importer (host country, receiving country, country of immigration) breaks down.

---

8. The scholarship on this dimension is voluminous. See inter alia Wayne Cornelius, "Spain: The Uneasy Transition from Labor Exporter to Labor Importer," in *Controlling Immigration: A Global Perspective*, eds. Wayne Cornelius, Philip Martin, and James Hollifield, 2nd ed. (Stanford, CA: Stanford University Press, 2004); Christian Joppke, *Immigration and the Nation-State: The United States, Germany, and Great Britain* (New York: Oxford University Press, 1999); Rogers Brubaker, "Immigration, Citizenship, and the Nation-Sate in France and Germany: A Comparative Analysis," *International Sociology* 5: 4 (December 1990), 379–407; and Aristide R. Zolberg, "Patterns of International Migration Policy: A Diachronic Comparison," in *Minorities: Community and Identity*, ed. Charles Fried (Berlin: Springer-Verlag, 1983), 229–246.

9. Laurie Brand, *Citizens Abroad: Emigration and the State in the Middle East and North Africa* (New York: Cambridge University Press, 2006); Natasha Iskander, *Creative State: Forty Years of Migration and Development Policy in Morocco and Mexico* (Ithaca, NY: Cornell University Press, 2010); and Jean-Pierre Cassarino, "Theorising Return Migration: The Conceptual Approach to Return Migrants Revisited," *IJMS: International Journal on Multicultural Societies* 6: 2 (2004), 253–279.

10. Douglas Massey et al. "Theories of International Migration: A Review and Appraisal," *Population and Development Review* 19: 3 (1993), 431–466."

11. Franck Düvall, *Crossing the Fringes of Europe: Transit Migration in the EU's Neighbourhood* (Centre on Migration, Policy and Society: University of Oxford Working Paper No. 33, 2006).

Five attributes characterize transit states.[12] First, they border an advanced-industrialized country (or grouping of countries) or offer reasonable access to it. Neither Tunisia nor Libya borders any EU member, but from them, migrants can access Malta, Sicily, or Lampedusa by boat. Morocco, Mexico, and Turkey obviously border Spain, the United States, and Greece, respectively. Kimball characterizes transit states as "stand[ing] precisely at the geographic crossroads of the first and third worlds."[13] Yet anachronistic Cold War categories such as "first world" and "third world" may be too stark. Indeed, in the cases of Mexico or Turkey, for example, the levels of development (using per capita income as an imperfect proxy) show that they are not subaltern, but rather strategic, emerging middle-income economies. Mexico is a member of NAFTA, while Turkey is a dynamic economy in its own right. Kimball also expands transit states to include a larger number of countries, such as EU members Poland, Czech Republic, Hungary, and Slovakia. This is too broad, given their position in the EU political space. To be fair, however, the countries joined the Schengen Area in 2007, after Kimball's study was completed.

Second, transit states themselves are invariably countries of emigration, with many citizens who live abroad in immigrant communities or aspire to emigrate. The latter are not transiting in the manner that irregular migrants traveling through the country are, but they are, in a different sense, in transit. And the movements of these citizens affect governance and the character of the state as it seeks to extend its purview beyond ostensible borders in order to govern its citizens abroad and engage other countries in negotiations where the interests of its émigrés are often part of a basket of diplomatic issues. Above all, transit states are keen to encourage repatriation of monies, as well as the regular return of émigrés during vacations, contributions to political campaigns, and voting in elections.

Third, obviously, transit states have in their territory non-nationals from a sending country or countries en route to an advanced-industrialized country. The ultimate destination of these migrants is heavily fortified, however. So they congregate in pockets of communities and in some instances are able to obtain work, often in the informal sector. From this

---

12. Ann Kimball, *The Transit State: A Comparative Analysis of Mexican and Moroccan Immigration Policies* (Center for Iberian and Latin American Studies and Center for Comparative Immigration Studies, University of California-San Diego, 2007); and Zeynap Sahin, "Policy Changes in the Immigration Controls of States after 1990s: The Case of Turkey" (International Studies Association Annual Meeting, New York, March 23, 2009).

13. Kimball, *The Transit State: A Comparative Analysis of Mexican and Moroccan Immigration Policies*, 3.

staging point, they endeavor to secure the services of smugglers, sometimes falling prey to traffickers.[14] For transit countries, the presence of undocumented migrants can be enormously challenging. They are often struggling with low levels of employment to begin with, and with domestic populations confronting anxieties similar to their advanced-industrialized counterparts. Immigrants, therefore, can encounter intense anti-immigration sentiment. The immigrant communities within transit states experience, at best, uneven integration into society. Transit states' ability to fully integrate immigrants is constrained by a lack of resources, a lack of political will, and the sense that the migrants are transitory to begin with.

Fourth, transit states must act vigorously in crafting their own immigration controls. They pursue this because of their own perceived interests, in collaboration with the advanced-industrialized countries that are the destinations of aspiring migrants. States fund their security apparatuses via both their own national budget and—more to the point—crucial foreign assistance. They participate in control efforts with full vim. Although respective interior ministries retain their national identities, in recent decades there has been a deepening transnationalization of interior ministries wherein cooperation concerning borders and immigration flows takes place at the highest levels. This is not to suggest that Tunisian and Italian authorities, for example, are always on the same page or that they lose sight of their own national identity. U.S. and Mexican officials, given their different positions within the international arena, certainly have different visions of appropriate migration controls. It is in the small differences, however, that the fundamental congruence of security endeavors is perceived in the "legitimate use of authority." Interior ministries and other law enforcement agencies cooperate across borders to share intelligence, coordinate efforts, and exchange fungible funds. Often, from the perspective of observers within a transit state, their country is serving as a gendarme or border guard for the destination country.[15] Whether the beat cop is doing the dirty work of a police commissioner may be beside the point: both are on the same force.

Some, but not all, transit states participate in frameworks in which they serve as an extension of North Atlantic security interests engaged in

14. Rey Koslowski, "The Mobility Money Can Buy: Human Smuggling and Border Control in the European Union," in *The Wall around the West: State Borders and Immigration Controls in North America and Europe*, eds. Peter Andreas and Timothy Snyder. (Lanham, MD: Rowman & Littlefield, 2000), 203–218.

15. Abdelkrim Belguendouz, *Le Maroc coupable d'émigration et le transit vers l'Europe* (Kénitra, Morocco: Boukili Impression, 2000).

counterterrorism efforts. This certainly is the case with respect to Mexico in the NAFTA context, as well as Morocco and Tunisia. Morocco is a non-NATO ally of the United States, and both Morocco and Tunisia have played roles as countries of rendition for the United States. As detailed below, Morocco, Tunisia, and, until January 2011, Libya also participate in NATO and EU efforts in the Mediterranean to stop irregular immigration.

Fifth, and perhaps most crucially, transit states often demonstrate an eagerness to participate in border control efforts as a way of enhancing their sovereignty claims and their credibility as a trusted diplomatic partner. By participating vigorously in regional groupings such as the EU or NAFTA, transit states can use migration controls to "reborder" their country. Spain sought to reborder itself after Franco's death in 1975, after the democratic transition in the late '70s, and especially after the accession to the European Economic Community (EEC) in 1986.[16] Yet North Atlantic states' externalization of border controls onto transit states has resulted in similar reborderings farther to the south. Mexico's efforts vis-à-vis Guatemala in the '90s and '00s can also be viewed as a means of strengthening its membership within NAFTA. Calls to rejuvenate these efforts were made after September 11.[17] One might also argue that rebordering is a way of cementing Mexico's sovereignty claims to Chiapas, which the Mexican government took from Guatemala in the nineteenth century after losing territory to the United States. Turkey's efforts in its southeast need to be seen in the context of Kurdish politics and relations with neighboring Iran, Syria, and Iraq. And, as discussed below, Rabat's claims to sovereignty over Western Sahara (and Ceuta and Melilla) are *always* front and center in Morocco's territorial efforts.

Thus, sovereignty assertions have to be examined in the context of state-building processes. As Kimball writes:

> Transit states are being courted, financially enticed, and diplomatically pressured to control their borders and detain transit migrants. . . . [B]y continuing to strengthen their relationships with [North Atlantic countries], transit states strategically align themselves with the developed North and distance themselves from the undeveloped South.[18]

16. Peter Andreas, *Border Games: Policing the US-Mexico Divide* (Ithaca, NY: Cornell University Press, 2000).

17. Clyde Hufbauer and Gustavo Vega-Cánovas, "Whither NAFTA: A Common Frontier?" in *The Rebordering of North America: Integration and Exclusion in a New Security Context*, eds. Peter Andreas and Thomas Biersteker (New York: Routledge, 2003), 128–152.

18. Kimball, *The Transit State: A Comparative Analysis of Mexican and Moroccan Immigration Policies*, 40.

This strategy is crucial. A common assumption is that middle-income countries are ossified, rooted in tradition. One often sees this in analyses of "developing" countries with emphases on "traditional" political structures, "challenges to modernization," "development strategies," and so on. Still, what we have seen in recent decades is tremendous dynamism in transit states vis-à-vis globalization, sovereignty norms, foreign affairs, and domestic governance. Morocco typifies this change.

MOROCCO AS A TRANSIT STATE: THE POLITICAL ECONOMY OF EMIGRATION

In order to appreciate Morocco's newfound role as a transit state, it is essential to emphasize its dual role as a country of emigration. The story of Morocco's evolution during the twentieth century into one of the world's foremost emigration countries is relatively well known.[19] During the interwar years, Moroccan labor in Europe was part of French colonial strategies in the Maghreb and a means of satisfying labor demands in the *métropole*. Although it slowed in the immediate aftermath of World War I, during the late '20s, French economic dynamism contributed to a resumption of Moroccan migration in the '20s and '30s.[20] Moroccan soldiers, known as *goumiers*, fought on behalf of the Allies during World War II. Bouachareb's *Indigènes*, a 2006 film about Algerian soldiers, illuminated the harsh discrimination Maghrebi soldiers experienced as they fought on behalf of the colonial power.

In the aftermath of World War II, immigrant manpower was crucial during "les trente glorieuses," the 30 years of postwar economic dynamism that stretched from 1945 until the early '70s. Moroccan migration during the '50s was important but did not reach significant levels until after Algeria achieved independence in 1962. During the '50s, Italian and Iberian workers fulfilled Europe's labor shortage. It was not until the '60s that Europe established guest worker programs with countries south of

19. Hein de Haas, *Morocco's Migration Transition: Trends, Determinants and Future Scenarios* (Nijmegen, The Netherlands: Centre for International Development Issues, Radboud University, 2005); Hein de Haas, "Morocco's Migration Experience: A Transitional Perspective," *International Migration* 45: 4 (2007), 39–70; Brand, *Citizens Abroad: Emigration and the State in the Middle East and North Africa*; Iskander, *Creative State: Forty Years of Migration and Development Policy in Morocco and Mexico*; and Jørgen Carling, "Unauthorized Migration from Africa to Spain," *International Migration* 45: 4–37 (2007), 3–37.

20. de Haas, "Morocco's Migration Experience: A Transitional Perspective."

the Mediterranean. Morocco, for its part, signed bilateral recruitment programs with West Germany (May 1963), France (June 1963), Belgium (February 1964), and Holland (May 1969).[21] Of the six members of the EEC at the time, only Luxembourg and Italy did not ink formal agreements. This is likely because of Luxembourg's small size and the fact that Italy's economy lacked dynamism, was a country of emigration in its own right, and already had informal Tunisian immigration. In the aftermath of guest worker programs, Moroccan emigration increased dramatically, gradually transforming into permanent settlement in Europe. In France alone, by 1970 there were roughly 10,000 Moroccans, joining 600,000 Algerians, 90,000 Tunisians, and 250,000 people from Guadeloupe, Martinique, and Réunion.[22]

Initially, the migration was largely male. Increasingly, however, it became more feminized as settlement and family reunification took place. Migrants occupied the bottom of the labor market and were often segregated in *bidonvilles* (shanty towns) and *banlieues* (suburbs). In the early stages, as Belguendouz has demonstrated, the bulk of immigration in the 20-year period from 1946 to 1966 was illegal; he puts the figure at 72 percent.[23] This runs counter to assumptions, especially given the official nature of the manpower accords signed, but the general dynamic was for a migrant to arrive as a tourist, find work, and remain. Return migration and resettlement in Morocco was rather regular, but in recent decades this has slowed considerably. Return migration among Moroccans has been among the lowest of immigrant groups in Europe, in part because of the progressive tightening of immigration policies in Europe and the deepening of the Schengen regime. With the Maastricht Treaty in 1992 and, especially, the Amsterdam Treaty, which entered into force in 1999, the Schengen Area now includes 25 European countries. Most, but not all, are members of the EU; some EU members are not full members. Additionally, many Moroccans have pursued naturalization in Schengen countries. In the '90s, 430,000 Moroccans obtained legal naturalization in Belgium, Denmark, France, Italy, Holland, and Norway. By the '90s there was also an increase in undocumented migration to Italy and Spain. In contrast to earlier guest worker migration, a considerable portion of this irregular flow was female,

21. Abdelkrim Belguendouz, *Le Maroc et la migration irrégulière: Une analyse sociopolitique* (Florence, Italy: Institut universitaire européen: Robert Schuman Centre for Advanced Studies, 2009), 2.

22. Stephen Castles and Mark Miller, *The Age of Migration: International Population Movements in the Modern World*, 4th ed. (New York: Palgrave Macmillan, 2010), 74.

23. Belguendouz, *Le Maroc et la migration irrégulière: Une analyse sociopolitique*, 2.

Table 4.1 COUNTRY OF RESIDENCE FOR MOROCCANS ABROAD

Country	Number (Percentage)
France	782,449 (28%)
Spain	691,848 (25%)
Italy	289,798 (10%)
Israel	222,222 (8%)
Holland	158,011(6%)
Germany	110,370 (4%)
Belgium	76,466 (3%)
United States	44,667 (2%)
Canada	27,816 (1%)
United Kingdom	13,725 (-%)
Total Abroad	2,718,665
Total Population	30,900,000

with women working as nannies, domestic workers, and cleaners, and in light manufactures.[24] De Haas estimates that the number of Moroccans officially living in Europe has increased ninefold, from 300,000 in 1972 to 2.6 million. The World Bank places the number of Moroccans living legally abroad at 2.7 million, or 8.6 percent of its 30 million people.[25] Table 4.1 shows the distribution by country in 2007.[26]

The reasons for this development are manifold. The primary impetus has been changes in and demand by the European political economy. The demographic profile of Europe has changed, with an aging, more affluent, and better-educated population emerging in recent decades. As a result, the demand for the flexible wage structure provided by an immigrant population has increased; by paying immigrants less, wages in the labor market are kept lower. As in other immigration contexts, much of the labor market is segmented according to race and national origin, with different

24. For Morocco and Spain see Natalia Ribas-Mateos, "Female Birds of Passage: Leaving and Settling in Spain," in *Gender and Migration in Southern Europe: Women on the Move*, eds. Floya Anthias and Gabriella Lazaridis (New York: Berg, 2000), 173–197. For a recent analysis of gender and the impact on society of immigration, see Moha Ennaji and Fatimi Sadiqi, *Migration and Gender in Morocco: The Impact of Migration on the Women Left Behind* (Trenton, NJ: Red Sea Press, 2008). For a broader discussion see Barbara Ehrenreich and Arlie Hochschild, eds., *Global Woman: Nannies, Maids and Sex Workers in the New Economy* (New York: Holt, 2004).
25. Dilip Ratha and Zhimei Xu, "World Bank Migration and Remittances Factbook," available at www.worldbank.org/prospects/migrationandremittances2009.
26. Calculated from data available at sitesources.worldbank.org.

groups working in different parts of the economy.[27] On the demand side of the equation, the "pull" for migrant labor is powerful. In 1975, as Spain's fascist regime died with Franco, Morocco and Spain shared roughly similar levels of economic development; today Morocco's GDP is roughly a sixth of Spain's.

At the same time, the "push" is significant, too. Moroccan emigration in the post-independence era has been encouraged by the Moroccan state for a variety of reasons. First, the monarchy has used emigration as a means of minimizing ethnic tension. This is especially evident in Tamazight, or Berber, regions (which include the Rif and High Atlas mountains and the Sous valley in the south). The monarchy's hold in these regions has long been tenuous, with uncertain allegiances toward the Palace. In the north of the country, for example, Rifi émigrés often worked in Ceuta and Melilla while endeavoring to work in Spain.[28]

Second, on a related note, emigration has been supported by the state as a way of transferring employment pressures abroad. To critics, this perpetuates the underdevelopment of the Moroccan economy and reinforces its subordinate position in the Euro-Maghrebi political economy.[29] To spell this out, there is considerable irony in Moroccan workers leaving rural areas because of the poor performance of the agricultural sector. To be sure, some of the sector's problems result from poor government policy, corruption, and chronic drought. But other, deeper constraints stemming from the colonial legacy, European protectionism, and free trade policies stifle agricultural production in Morocco (and other developing countries).[30] Moroccan farmers then emigrate to neighboring Spain to work as pickers in the Andalusian provinces of Huelva and Jaen or to the east in Murcia, harvesting olives, strawberries, and citrus that they could conceivably farm back home. Spain, after all, has a similar production profile and the same sunshine. However, since 1986, Spain has prospered

27. Carlota Solé and Sònia Parella, "The Labour Market and Racial Discrimination in Spain," *Journal of Ethnic and Migration Studies* 29: 1 (2003), 121–140.

28. David McMurray, *In and Out of Morocco: Smuggling and Migration in a Frontier Boomtown* (Minneapolis: University of Minnesota, 2001); and Gregory White, "La migración laboral Marroquí y los territorios Españoles de Ceuta y Melilla," *Revista internacional de sociología* 36 (September–December 2003), 135–168.

29. Aristide R. Zolberg, "Patterns of International Migration Policy: A Diachronic Comparison"; and Michael Piore, *Birds of Passage: Migrant Labor in Industrial Societies* (Princeton, NJ: Princeton University Press, 1979).

30. Will Swearingen, *Moroccan Mirages: Agrarian Dreams and Deceptions, 1912–1986* (Princeton, NJ: Princeton University Press, 1987); and Gregory White, *On the Outside of Europe Looking in: A Comparative Political Economy of Tunisia and Morocco* (Albany, NY: State University of New York Press, 2001).

because of its position within the EU's hyper-protective Common Agricultural Policy (CAP).

This dynamic abets the third reason for Morocco's reliance on emigration: the need to secure foreign exchange.[31] Remittances—monies sent back home—are a crucial source of financial capital. They dwarf direct foreign investment, earnings from other economic activities, or funding for development projects back home. According to the World Bank, Moroccans living legally abroad remitted US$5.7 billion in 2007, or 10 percent of the GDP for that year.[32] In rankings of total monies or remittances as a percentage of GDP, Morocco is near the top. For example, in terms of total monies, it is second to Egypt in the region; as a percentage of GDP, its remittances place it fourth, behind Jordan, Lebanon, and the West Bank/Gaza. Adding the undocumented migrants living abroad would make this figure significantly and incalculably higher.

Fourth, social capital emerged as migrants deployed networks developed overseas. In some instances, migrants return home and capitalize on the contacts made abroad, encountering challenging social dynamics in the process.[33] In other instances, they visit home on an annual or regular basis. The most prominent instance of this is the collective summertime vacation—*opération transit* or *opération marhaba*—when Moroccan migrants cross legally (heading southward) by ferry from Algeciras, Spain, or from Sète, France; they also arrive by plane at Casablanca's Mohammed V Airport. To return to the remittance point above, even if they do not return home physically, social capital is created when they send monies home for specific projects. Rural development projects have been funded by émigrés who want to help build infrastructure in their home community.

The state was laissez-faire vis-à-vis émigrés in the '60s and '70s, but no more. As detailed below, the state has responded with institutional reforms to address immigration. Morocco, thus, is an emigration country par excellence—an identity that has emerged only in recent decades. It has evolved in terms of its institutions and governance structures to capitalize on all of these dynamics.

Before turning to the policy response of a transit state, it is important to note the distinction—often fuzzy and hard to quantify—between legal and illegal migration. The distinction is certainly part of the political

31. Gregory White, "The Maghreb in the World's Political Economy," *Middle East Policy* 14: 4 (Winter 2007), 42–54.

32. Ratha and Xu, "Migration and Remittances Factbook."

33. Cassarino, "Theorising Return Migration: The Conceptual Approach to Return Migrants Revisited."

landscape. It is also the stuff of art, with novelists, musicians, and filmmakers plumbing the experience of *Hrig* or *Haragas*—people who "burn" borders by burning their documents to avoid being sent home. Writers such as the celebrated Tahar Ben Jelloun or newcomer Laila Lalami offer powerful depictions of emigrant experiences.[34] Lalami's remarkable *Hope and Other Dangerous Pursuits* is a poignant collection of interwoven short stories about migrants from Morocco and other countries seeking to cross the Strait of Gibraltar. Some of her characters appear to be climate-induced migrants. Jelloun's *Leaving Tangier* also interweaves immigrant stories about crushed hopes. Rock bands sing of youthful alienation and dreams of emigration. Casablanca-based Hoba Hoba Spirit mixes reggae, punk, and Gnawa; imagine Joe Strummer and Bob Marley jamming with Nass El Ghiwane or Jil Jillala—legendary Moroccan groups. André Téchiné and Faouzi Bensaïdi's *Loin* (2001), Douad Aoulad-Syad's *Tarfaya* (2004), and José Luis Tirado's documentary *Latitude 36* (2004) are outstanding film depictions of politics at the Moroccan-Spanish border. *Loin* (or "Far") skillfully portrays the human drama in Tangiers, where the trade of textiles between Morocco and Europe is mixed with the illicit trade of drugs and migrants. Finally, artists have crafted exhibits devoted to the flux of peoples at the Spanish-Moroccan border. Isaac Julian's 2007 multimedia installation *Western Union: Small Boats*, with its stark images of boats and corpses splayed on beaches, has been displayed in Spain, Portugal, and, most recently, Marrakech.[35]

THE EXPERIENCE WITH IMMIGRATION

Despite Morocco's oft-noted position at the crossroads of the Mediterranean, if its experience with emigration is a post–World War II phenomenon, its story as a country of immigration is even more recent. Morocco has rarely been understood as a country of significant immigration, with communities of people from elsewhere settling there. Of course, one must not neglect the history of slavery as contributing to the complexity of

34. Laila Lalami, *Hope and Other Dangerous Pursuits* (Chapel Hill, NC: Algonquin Books, 2005); and Tahar Ben Jelloun, *Leaving Tangier: A Novel*, trans. Linda Coverdale (New York: Penguin, 2009).

35. See www.isaacjulien.com. See also Holiday Powers, "The Challenges of Maintaining Local Identity in International Biennale Exhibitions: Lessons from the 3rd AiM Arts in Marrakech Biennale," paper presented at the 2010 AIMS Conference, June 27, 2010, University of Oran, Algeria.

Moroccan society; such trade continued well into the twentieth century.[36] Northwest Africa also hosted nomadic populations for millennia, given its central role in the trans-Sahara trade system. Ruefully, one should also mention French and Spanish colonial "immigration" into Morocco; the resulting communities were not on the scale of the French settlement in Algeria after 1830, but they were significant and remain so. Nonetheless, Morocco did not experience immigration on the order experienced by advanced-industrialized economies—or oil economies such as Algeria, Libya, or the Gulf states.

This began to change in the late '80s. As Belguendouz noted, *pateras* illegally crossing the Strait of Gibraltar were intercepted for the first time beginning in the '90s.[37] The arrival of immigrants corresponded roughly to the country's adoption of neoliberal economic reforms in exchange for structural adjustment package loans from the World Bank and IMF, and the gradual emergence of the economy from statist tutelage. Another factor was the southern expansion of the EEC in 1986 to include Spain and Portugal, not to mention the steady tightening of European migration controls. This, coupled with the political instability and poor performance of sub-Saharan African economies throughout the '70s and '80s, contributed to flows north into Morocco. Like Moroccan emigrants, transit migrants responded to the pull stimuli for labor in an aging European political economy. And they share many of the same motivations: Europe's employment opportunities, better lifestyles, and remittances.

Above all, it is important to bear in mind what de Haas has accurately termed the "myth of the invasion."[38] As he argues, it overplays the scope and dimension of migration to Europe via Morocco and other Maghrebi countries. It also ignores the structural demand for cheap labor in informal sectors of the economy in Europe and, increasingly, the Maghreb. According to de Haas's rough estimation, 120,000 people enter the entire Maghreb each year. This number is significant and, again, will likely grow in the decades to come. Nevertheless, it is certainly not the horde so often offered by the North Atlantic media or by security-minded analysts.

Specifically with respect to Morocco, according to the Moroccan Association for the Research and Study of Migration (AMERM), in 2007 6,500

36. Mohammed Ennaji, *Soldats, domestiques et concubines: L'esclavage au Maroc au XIXe siècle* (Casablanca: Editions EDDIF, 1997).

37. Belguendouz, *Le Maroc et la migration irrégulière: Une analyse sociopolitique*, 4.

38. Hein de Haas, *The Myth of Invasion: Irregular Migration from West Africa to the Maghreb and the European Union* (Oxford, England: International Migration Institute, 2007).

sub-Saharans lived in the large cities—Rabat, Casablanca, Oujda, Laây-oune, and Tangiers.[39] The European Commission funded a report in 2009 that estimated Morocco's population of irregular migrants as 10,000, and the IOM has estimated between 10,000 and 20,000.[40] These are significant numbers to be sure, and it would hardly be appropriate to dismiss them as irrelevant. The thousands of sub-Saharans in the major cities congregate in quarters and have emerged as an important presence. In Rabat, for example, migrants have settled in poor, working-class quarters such as Taka-doum and Yacoub Mansour. These neighborhoods emerged only in the '70s as part of a broader process of rapid urbanization in the country.

One significant driver of sub-Saharan Africans into Morocco was the Ivory Coast civil war that began in 2002 and intensified in 2004, with crucial implications for neighboring countries. According to AMERM, the sub-Saharan population in Morocco is divided by country of origin as follows: Nigerians (15.7 percent), Malians (13.1 percent), Senegalese (12.8 percent), Congolese (10.4 percent), Ivoirians (9.2 percent), Guineans (7.3 percent), Cameroonians (7 percent), Gambians (4.6 percent), Ghanaians (4.5 percent), Liberians (3.8 percent), and Sierra Leoneans (3.1 percent).[41]

The routes taken by migrants into Morocco are informal and hard to specify. Cartographic depictions abound and offer a valuable epistemological tool for understanding these routes. At the same time, maps tend to freeze one's understanding both spatially and temporally. Further, they may intentionally or unwittingly amplify the notion of multitudes converging on a bottleneck such as the Strait of Gibraltar or Lampedusa. One map provided by the EU's Frontex and available on the BBC's website highlights migrant routes in red.[42] The Vienna-based International Centre for Migration Policy Development (ICMPD) developed a similar map in 2007.[43] It too is based on information developed in collaboration from Frontex and Europol. ICMPD, the UNHCR, Frontex, and Europol have been at the forefront of efforts to institute collaboration between European and Maghrebi governments on irregular migration, an example of the epistemic community forging interstate networks of security. A third map that is illustrative was developed by de Haas. It is superior, in part because it eschews the

39. Laura Feliu Martínez, "Les migrations en transit au Maroc: Attitudes et comportement de la société civile face au phénomène," *L'Année du Maghreb* 5 (2009), 345.
40. Emily Pickerill, "Informal and Entrepreneurial Strategies among Sub-Saharan Migrants in Morocco," *Journal of North African Studies*, (forthcoming).
41. Feliu Martínez, "Les migrations en transit au Maroc: Attitudes et comportement de la société civile face au phénomène," 344.
42. Available at news.bbc.co.uk/2/hi/europe/6228236.stm.
43. Available at mtm-imap.net.

alarmist red routes of the Frontex/BBC map; it is also comprehensive in its coverage of the region.[44] By way of illustration, map 4.1 combines the attributes of these maps, namely the ostensible directionality of the migration routes, as well as their trans-regional character.

While such maps are invaluable, the degree to which they emphasize political boundaries reveals an important set of assumptions about borders and their permanency and legitimacy. International cooperation on migrants presumes sovereignty and, perhaps, the ability to control the borders. Most of all, they underestimate the historical mutability of borders. As Charef writes:

> At one time in the history of Morocco, it was possible to travel from the banks of the Senegal river to the north of Spain along the same road within the borders of the same Kingdom! Glorious and blessed times according to many chroniclers. . . . [W]e find ourselves wondering about the revival of caravan routes by these latter-day nomads forced by circumstance into the role of clandestine migrants! It is a fact that many Sub-Saharan nations are following the old routes to the Mediterranean coast (their aim and objective), making the same journey their forbears made for different reasons.[45]

Today, to be sure, many borders in the Maghrebi, Saharan, Sahelian, and sub-Saharan regions of Africa are not patrolled or even marked. Migrants who do brave the journey north encounter brutal conditions as they cross enormous tracts of land. Often they do not meet officials.[46] For that matter, official documents are often nonexistent; if they do exist, and a border officer is encountered, passage can be secured for a bribe.

It is important to also stress that many migrants do not traverse a transit country but instead remain once they arrive.[47] This has been the case with Turkey, Libya, and Mauritania, and is increasingly the case for Morocco. As is well known, cities become hubs for global migration, so Nouakchott, Tripoli, Casablanca, and Algiers have significant numbers of

---

44. de Haas, *Myth of Invasion*, 21.

45. Mohamed Charef, "Geographical Situation as a Facilitator of Irregular Migration in Transit Countries—the Case of Tangier" (Istanbul, Council of Europe, September 30–October 1, 2004), 47.

46. Michael Collyer, *States of Insecurity: Consequences of Saharan Transit Migration* (Oxford, England: University of Oxford Centre on Migration, Policy and Society, 2006), WP-06-31.

47. Hein de Haas, *Irregular Migration from West Africa to the Maghreb and the European Union: An Overview of Recent Trends* (Geneva: International Organization for Migration, 2005), 64.

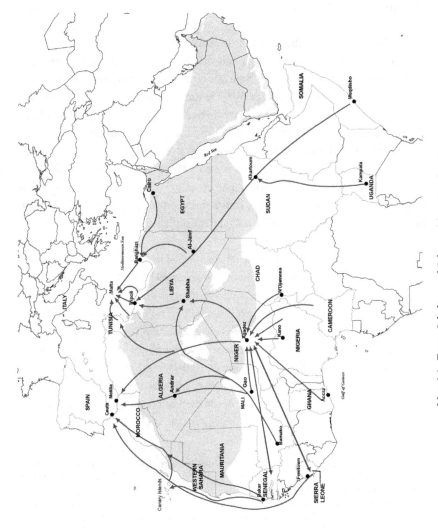

Map 4.1   Trans-Sahelian and Saharan Migration Routes

migrants who have settled to work in petty trade, service sectors, and construction.[48] Although Maghrebi countries are not doing well vis-à-vis Europe, as Baldwin-Edwards notes, "In comparison with the majority of African countries even North Africa looks prosperous, and it is moreover closer to Europe and therefore a stepping stone to a better life."[49] Whether this portends the beginning of a full settlement process remains to be seen.[50]

For the time being, then, to cast Morocco as a new country of immigration, as some analysts and European officials have suggested, is fraught with implications. Immigration typically connotes a settlement process: the migrant wants to arrive and stay. Yet it is hard to imagine calling Morocco a destination country or a host country when so many of its migrants have distant aspirations and yet are blocked and caught in between.[51]

Finally, official news media emphasize the ostensible problems that Sahelian and sub-Saharan Africans pose to Morocco. Such sources include the Maghreb Arab Press (MAP) Agency, television stations like 2M and RTM, and radio stations like the Tangiers-based Medi 1. The MAP Agency, for example, published a 2009 analysis in English:

> At a stone's throw from Europe and due to a gained reputation of transit to the Continent, Morocco faces increasing challenges to deal with the thorny issue of several hundreds of sub-Saharan would-be illegals who cross to the country. While a few succeed the crossing, a lot remain stranded in a country of handling the continuing arrivals. Media reports suggest that most sub-Saharan migrants leave their native countries with the hope of driving better lives for themselves and their families. . . . Morocco is 15 km (9.3 miles) from Spain. It is close to the Canary Islands and has two Spain-occupied cities, Sebta and Melilia, which makes is [sic] ideally the last-but-one destination before the final destination.[52]

48. Saskia Sassen, "Whose City Is It? Globalization and the Formation of New Claims," in *Globalization and Its Discontents: Essays on the New Mobility of People and Money* (New York: Free Press, 1998).

49. Martin Baldwin-Edwards, "Between a Rock and a Hard Place: North Africa as a Region of Emigration," *Review of African Political Economy* 33: 108 (2006), 322.

50. de Haas, *Irregular Migration from West Africa to the Maghreb and the European Union: An Overview of Recent Trends*, 64.

51. Jérôme Valluy, "Aux marches de l'Europe: des 'pay-camps': La transformation des pays de transit en pays d'immigration forcée (observations à partir de l'exemple marocain," in *Immigration sur emigration: Le Maghreb à l'épreuve des migrations subsahariennes*, ed., Ali Bensaâd (Paris: Karthala, 2008), 325–342.

52. Abdallah M'channa, "Sub-Saharans in Morocco: The Mid Path to El Dorado," *Agence Maghreb Arabe Presse*, May 15, 2009, available at www.map.ma/eng/sections/general/sub-saharans_in_moro/view.

To its credit, the document referred to "several hundreds" of immigrants rather than thousands or tens of thousands. Interestingly, it also referred to Sebta and Melilia as "Spain-occupied," a common trope for Moroccan nationalism. The bottom line is that the official media itself is emphasizing the "increasing challenges" posed by immigration. In some instances, opinion pieces cross the line. In September 2005, a Tangier newspaper, *al-Shamaal*, referred to sub-Saharan Africans as "black locusts" invading the country.[53] Moroccan authorities promptly banned the paper for the racist transgression, but the sentiment was revealing.

STATE RESPONSE—POLICY MEASURES

While it is difficult if not impossible to know the full scope of immigration into Morocco—or the number of migrants who, in turn, attempt to travel to Europe—what is discernable is the response of Moroccan society and, more central to this discussion, the state. At the societal level, the response to immigrants has been grudging and sometimes inhospitable, as it often is around the world. Given the poor performance of the Moroccan economy, there is hardly an absorptive capacity for employment. As noted above, the media often covers immigration to the country, as well as the emigration (legal and undocumented) of its own citizens, in a sober fashion. It certainly covers official collaboration between the Moroccan state and its European counterparts, routinely announcing meetings between Moroccan security officials and NATO and interior officials from other countries.

Moroccan society is relatively weak in comparison to the power of the monarchy and the *Makhzen*, the powerful central governing institution arrayed around the palace.[54] In North Atlantic host countries, the state has to engage with powerful interest groups, be they groups representing

53. Elie Goldschmidt, "Storming the Fences: Morocco and Europe's Anti-Migration Policy," *Middle East Report Online* 239 (2006), available at http://www.merip.org/mer/mer239/goldschmidt.html.

54. See inter alia Azzedine Layachi, "State-Society Relations and Change in Morocco," in *Economic Crisis and Political Change in North Africa*, ed. Layachi (Westport, CT: Praeger, 1998), 89–106; Guilain Denoeux and Abdeslam Maghraoui, "King Hassan's Strategy of Political Dualism," *Middle East Policy* 5: 4 (January 1998), 104–130; Abdeslam Maghraoui, "From Symbolic Legitimacy to Democratic Legitimacy: Monarchic Rule and Political Reform in Morocco," *Journal of Democracy* 12: 1 (2001), 73–86; John Waterbury, *The Commander of the Faithful: The Moroccan Political Elite—A Study in Segmented Politics* (New York: Columbia University Press, 1970); and Abdellah Hammoudi, *Master and Disciple: The Cultural Foundations of Moroccan Authoritarianism* (Chicago: Chicago University Press, 1997).

employers seeking less restrictive policies, groups on the xenophobic right protesting immigration, or groups defending immigrants' interests.[55] In the context of a patrimonial state, however, social groups have little heft. Thus, the state's ability to enact top-down policies concerning transit migration is much greater. NGOs in Morocco to protect African immigrants and refugees are active, yet their ability to influence protection for immigrant populations is limited. The Moroccan Organization for Human Rights (OMDH, Organisation Marocaine des Droits Humains), the Anti-Racist Group for the Accompaniment and Defense of Foreigners and Immigrants (GADEM, Groupe anti-raciste d'accompagnement et de défense des étrangers et migrants), and Friends and Families of Victims of Clandestine Immigration (AFVIC, Amis et familles des victimes de l'immigration clandestine) are significant actors that have grown in importance.[56] Nonetheless, they remain inhibited.

Morocco is conventionally portrayed as rooted in tradition and unchanging. This is especially the case in efforts to brand the country for tourists seeking to experience a timeless, authentic Morocco. The *Makhzen* is often cast as fixed and immutable by journalists, tour operators, and social scientists. The monarchy derives authority and allegiance, or *bay'a*, from the king's position as a descendent of the Prophet and Commander of the Faithful.[57] Yet, in fact, the state is intensely dynamic. While its response to internal and external forces is not always adept or rapid, it does evidence significant adaptive capacity. This is especially the case with respect to migration policy. As Morocco's experience with emigration and immigration has evolved over the decades, the state's response has evolved as well.

Initially, to return to the dynamics associated with emigration, in the early years of independence and during the period of the bilateral accords signed with European countries, the state was very much involved through passport issuance. This was often brokered and carefully controlled by local elites to secure privilege. In the '70s, the establishment of Amicales or Wiladiat (government-controlled migrant associations based overseas), along with embassies, consulates, and mosques, helped control migrant populations. Their task was often one of surveillance and the extension of

55. G. J. Borjas, "The New Economics of Immigration: Affluent Americans Gain, Poor Americans Lose," *Atlantic Monthly*, November 1996.

56. For a "mapping" of NGOs working on behalf of transit migrants, see Feliu Martínez, "Les migrations en transit au Maroc: Attitudes et comportement de la société civile face au phénomène."

57. Maghraoui, "From Symbolic Legitimacy to Democratic Legitimacy: Monarchic Rule and Political Reform in Morocco"; and Waterbury, *The Commander of the Faithful: The Moroccan Political Elite-A Study in Segmented Politics.*

state sovereignty beyond the border. This deterritorialization of the state has been evident in other realms as well, including, in the '90s and '00s, Internet control. In Tunisia, for example, the Ben Ali regime sought to control citizenship and dissent within and beyond its borders by disrupting Internet backbones and nodes and monitoring traffic.

In 1990, as the nature of the émigré community abroad changed, becoming less male and more inclined to reside for a long time abroad and with reunified families, the Amicales became less viable. The *Makhzen* created a Ministry of the Moroccan Community Abroad (MMCA) that year to provide services for Moroccan residents abroad. Designed to bridge the dossiers of the Ministries of Labor and Foreign Affairs, the MMCA devoted itself to understanding the lay of the land in respective destination countries as well as to dealing with legal questions, data, and media relations between émigrés and the home country. In time, however, the MMCA declined and was downgraded to a subministry because of bureaucratic infighting. Ultimately, in 1997, its dossier was reabsorbed into a reinvigorated Ministry of Foreign Affairs.[58]

On a parallel track, in 1990 the government created a parastatal organization: the Hassan II Foundation for Moroccans Residing Overseas (FHII, Fondation Hassan II pour les Marocains Résident à l'Étranger). This, too, struggled throughout the '90s in fulfilling its mandate to provide language instruction to children of Moroccans abroad, vacation camps for émigré children during the summer, and financial assistance. In 2000, the Mohammed V Foundation for Solidarity (Fondation Mohammed V pour la Solidarité) took over responsibility for the summertime *Opération Marhaba*. And in the early '00s, the Moroccan state established Al Maghrebia, a satellite channel presented in Arabic, French, Tamazight, and Spanish dedicated to the Moroccan community abroad. The channel is available at a fascinating website, www.snrt.ma (Société nationale de radiodiffusion et de télévision). The state has fully recognized that émigrés abroad are inclined to stay for long periods and that support for their activities is necessary.

CONTROLLING TRANSIT MIGRATION

To return to the issue of transit migration and immigration pressures from the south, official efforts by the Moroccan state effectively began with a 1991 bilateral treaty signed with Spain. The treaty's title, "The Treaty of Friendship, Good-Neighborliness and Cooperation," was rather droll, as

58. Brand, *Citizens Abroad: Emigration and the State in the Middle East and North Africa*.

the two antagonists pronounced perhaps too loudly that they really did like each other. Tensions between Rabat and Madrid had been in place since Spain's democratic transition in the late '70s and continued well into the '00s.[59] A close reading of the treaty reveals that it did not mention irregular migration from Morocco to Spain.[60] Nonetheless, the treaty provided the context for a grudging rapprochement, including a visit to Ceuta by Hassan's interior minister, Driss Basri. For years, Basri was a notorious strongman in charge of a bureaucracy that, in the Moroccan context, was a superministry combining police, information, and intelligence powers. Along with Hassan, he was the face of the ignoble "years of lead"—the height of Morocco's civil rights abuses.[61] In making an official visit to Ceuta, the interior minister was effectively acknowledging Spanish sovereignty, much to the consternation of hard-line nationalists.

The Moroccan state furthered its efforts to control immigration in the south, or at least proclaimed its efforts in this regard, throughout the '90s by participating in multilateral frameworks and collaborating with Spanish and European officials. In 1992, Morocco and Spain signed an accord stipulating that Morocco accepted responsibility for readmitting citizens and third-country nationals arriving illegally on Spanish territory. However, this was not well received by the Moroccan media, and the law was not well applied by the Moroccan authorities.[62] In June 1993, Morocco became the second country after Egypt to ratify the 1990 UN Convention on the Protection of the Rights of all Migrants and Members of Their Families. Pushed largely by countries in the global south, the signing was largely symbolic and had little effect.[63] Morocco also signed some of the International Labor Organization (ILO) and UN instruments.

59. Bernabé López-García, "Foreign Immigration Comes to Spain: The Case of the Moroccans," in *New European Identity and Citizenship*, eds. Rémy Leveau, Khadija Mohsen-Finan, and Catherine Withol de Wenden (Aldershot, UK: Ashgate, 2002), 49–68; Miguel Hernando de Larramendi, *Las relaciones con Marruecos tras los atentados del 11 de Marzo* (Madrid: Real Instituto Elcano de Estudios Internacionales y Estratégicas, 2004); and Richard Gillespie, *Spain and Morocco: Towards a Reform Agenda?* (Madrid: Fundación para las relaciones internacionales y el diálogo exterior, 2005).

60. Kingdom of Spain and Kingdom of Morocco, Treaty of Friendship, Good-Neighbourliness and Cooperation-Signed in Rabat, Morocco, no. 1717, I-29862 (July 4, 1991).

61. Susan Slyomovics, *The Performance of Human Rights in Morocco* (Philadelphia: University of Pennsylvania Press, 2005).

62. Lucile Barros, Mehdi Lahlou, Claire Escoffier, Pablo Pumares, and Paolo Ruspini, "L'immigration irrégulière subsaharienne à travers et vers le Maroc," *Cahiers de migrations internationales* 54F (Bureau International du Travail: Genève, 2002).

63. Khadija Elmadmad, *Migration irrégulière et migration illégale: L'exemple des migrants subsahariens au Maroc* (Florence, Italy: Robert Schuman Centre for Advanced Studies-European University Institute, 2008).

Also relevant in the evolution of Moroccan migration politics was the 1996 signing of a celebrated Association Agreement with the European Union.[64] The agreement was negotiated in the context of the Barcelona Process and treated a wide range of issues, including trade, financial assistance, technical assistance, and cultural exchange. Diplomacy took place in a controversial context. Points of contention included other difficult issues such as access to Moroccan fisheries, drug interdiction, surveillance against dissidents, trade, and energy. The final agreement also attended to the issues of Moroccans working in Europe, although Moroccans resented that the agreement lumped immigration with such problems as organized crime, drugs, and terrorism.[65]

The ascendance of Mohammed to the throne in July 1999 and, in turn, the September 11 terror attacks marked a crucial turning point within Morocco. On the one hand, Mohammed presided over a gradual opening of the political system, one grudgingly started by his father in the late '90s.[66] He dismissed Interior Minister Basri in September 1999 and began to open the dossiers on human rights abuses. Elections in September 2002 portended a grudging democratization as well. At the same time, September 11 had prompted a deepening of the security relationships between Morocco and its North Atlantic allies, with Morocco ultimately becoming a non-NATO ally in 2004.

At the Council of Europe's 2002 summit in Seville, the Spanish government proposed to sanction governments that did not participate sufficiently in the effort to reduce migratory flows. This was clearly directed at Morocco and was ultimately rejected after French pressure. Still, Morocco's efforts changed. In 2004, joint border patrols of Spanish and Moroccan boats took place between Laâyoune and Las Palmas in the Canary Islands.[67] Also notable is the Spanish government's adoption of an Africa Plan, ostensibly devoted to supporting African development but also focused on slowing the flow of African immigrants. In July 2008, joint commissions were created between the two governments.

In the realm of immigration control within Morocco, efforts continued in November 2003 with the passage of a new law, Dahir No. 1.03.196,

64. Text available at ec.europa.eu/external_relations/morocco/association_agreement/index_en.htm.

65. Belguendouz, *Le Maroc coupable d'emigration et de transit vers l'Europe.*

66. Gregory White, "'The End of the Era of Leniency' in Morocco? Mohammed VI's Halting Glasnost," in *North Africa: Politics, Religion and the Limits of Transformation,* eds. Y. Zoubir and H. Amirah-Fernández (London: Routledge, 2008), 90–108.

67. Feliu Martínez, "Les migrations en transit au Maroc: Attitudes et comportement de la société civile face au phénomène."

regulating the entry and residence of foreigners. It is more commonly known as Law 02-03. The law distinguishes between legal and illegal migrants and between migrant workers, refugees, and asylum seekers.[68] It is unprecedented in Morocco, mirroring immigration legislation in North Atlantic countries. The analog in the U.S. context would perhaps be the Clinton administration's IIRIRA in 1996. In Spain, it would be the Aznar government's passage of immigration laws in the '90s. One might be tempted to argue that Morocco's law brought it into the club of advanced-industrialized countries, with official policies discouraging immigration. Yet given Morocco's role as a transit state, the implications on the ground are very different.

In fact, during the same session in the Moroccan *majlis*, or parliament, a second law was passed concerning antiterrorism efforts. Throughout the period, a security preoccupation was at work. A key impetus for Law 02-03 was the May 16, 2003, terror attacks in Casablanca, in which 33 civilians and 12 bombers died in separate, coordinated bombings. The attackers were homegrown, Moroccan-born men, not immigrants, acting in the name of Al Qaeda. In Morocco, they were affiliated with the Salafia Jihadia group. Still, the state targeted immigrants as a source of instability. The attacks rocked the political landscape, affecting municipal elections that year as well as national elections in 2007. Moreover, they shaped the debate over the passage of the *Mudawwana*, the family status code that governs marriage law and women's rights. In an intriguing analysis, Salime argues that the May 16 bombings ironically abetted women's rights groups in pressuring the Palace to marginalize the legal Islamist party, the Justice and Development Party (PDJ), and to finalize the legislation in 2004.[69]

With the passage of Law 02-03, as Elmadmad argues, "Migration [in Morocco] is more and more tied to terrorism and criminality."[70] The law includes heavy sanctions against undocumented immigration and human smuggling but largely ignores migrants' rights. It established "waiting zones" and "retention centres."[71] And it set in motion events such as the forced movement of migrants in 2005 to the Algerian border near Oujda, discussed below. In passing the new law Morocco was bowing

---

68. Fatima Sadiqi, *Migration-Related Institutions and Policies in Morocco* (European University Institute, Florence, Italy: Euro-Mediterranean Consortium for Applied Research on International Migration [CARIM], 2004).

69. Zakia Salime, "The War on Terrorism: Appropriation and Subversion by Moroccan Women," *Signs: Journal of Women in Culture and Society* 33: 1 (2007), 1–24.

70. Elmadmad, *Migration irrégulière et migration illégale: L'exemple des migrants subsahariens au Maroc*, 6.

71. Pickerill, "Informal and Entrepreneurial Strategies among Sub-Saharan Migrants in Morocco."

to pressure from the EU as it embraced the role of Europe's "policeman" in North Africa.[72]

The law has contributed to the further evolution of governance within Morocco in that it established new state bureaucracies: the Directorate of Migration and Border Surveillance and a Migration Observatory, both within the Interior Ministry. Some of the apprehended immigrants are expulsed to countries of origin. In 2004, Nigeria accepted five planes with 1,700 Nigerians who had been residing illegally in Morocco.[73] Most undocumented migrants, however, appear to be either imprisoned or expulsed into the desert of Algeria. According to Doctors without Borders, illegal migrants often suffer violence at the hands of Moroccan security services.[74]

Morocco works to accentuate the differences between émigrés and immigrants on several levels. First, Moroccans have access to Ceuta/Melilla as day laborers under the Schengen agreement. A daily quota can pass through border posts. There are also holes in the fence that facilitate the informal/extralegal transport of goods. Second, Moroccans caught trying to cross to Spain are repatriated, whereas non-Moroccan nationals are processed. Third, within Europe there is a differential treatment (especially between September 2001 and Aznar's departure from office in March 2004); Morocco has consistently sought to protect its nationals' interests abroad.

Strictly speaking, as noted earlier, the state's concern for Moroccan emigration is not transit. But this is the point: the Moroccan government is urged by Spain and EU to help slow the emigration of its own population. And Morocco responds with a kind of caginess or even insolence. If it could control the emigration of its own citizens—*if it could*—it would not want to do so. So it turns to an arena where it can facilitate control: migrants seeking access to Europe through Morocco.

THE 2005 STORMING OF THE ENCLAVES
AND RECENT DEVELOPMENTS

In 2005, Spanish-Moroccan relations took an abrupt turn in terms of cooperation between the governments. In part, this was due to the Zapatero government's willingness to work with Morocco after it came to office in

---

72. de Haas, *Irregular Migration from West Africa to the Maghreb and the European Union: An Overview of Recent Trends*, 64.
73. Baldwin-Edwards, "Between a Rock and a Hard Place: North Africa as a Region of Emigration," 321.
74. Ibid.

March 2004, in sharp contrast to the Aznar government. In the aftermath of the March 11, 2004, terror attacks on the Atocha train station and the implication of Moroccans nationals, Rabat stepped up its cooperation in terrorism and security. In April 2004, Zapatero visited Rabat, and throughout the rest of the year and into the next, Spanish-Moroccan cooperation accelerated. Again, it is hard to find evidence of quid pro quo, but the signs of enhanced cooperation were abundant. In November 2004, the Spanish secretary of state visited Rabat and announced an additional supplementary aid budget of €950,000. In December 2004, the Spanish *Guardia Civil* began joint patrols with Morocco's Royal Police. In January 2005, King Juan Carlos publicly thanked Morocco for its cooperation in controlling immigration from sub-Saharan Africa. In February 2005, Morocco authorized the establishment of a regional office for the International Organization of Migration. And in the same month, nearly 500,000 undocumented workers were regularized in Spain, the bulk being Moroccan in origin.[75]

In September 2005, African migration to Europe achieved international prominence when hundreds of migrants stormed the Spanish presidios. Reports are ever murky, but it appears that migrants around Ceuta and Melilla coordinated their effort to rush the fences, throw ladders and blankets over the razor wire, and clamber into Spain. They were beaten back by Moroccan and Spanish officers. The official toll was 14 migrants shot, with many injured and thousands expelled from Morocco into Algeria.

Subsequently, in July 2006, Morocco trumpeted its crucial role in regional cooperation on illegal migration and hosted a Euro-African Ministerial Conference on Migration and Development. Undergirded by the Spanish and French governments, the resulting Rabat Declaration brokered multilateral cooperation on illegal migration.[76] The document reads well. Although it does not mention CIM explicitly, it touches on crucial aspects of supporting development and on the importance of properly managing immigration from sub-Saharan Africa. It also contains strong language on the importance of controlling borders. Throughout 2006 and into the latter part of the decade, Moroccan authorities stepped up their raids of migrants around cities such as Nador (near Melilla) and pursued deportations of migrants to Algeria.

In June 2009, in language unimaginable when Aznar was Spain's president, Immigration and Emigration State Secretary Consuelo Rumi

75. Valluy, "Aux marches de l'Europe: des 'pay-camps,'" 331.

76. Euro-African Ministerial Conference on Migration and Development, "Rabat Plan of Action of the Euro-African Ministerial Conference on Migration and Development" (Rabat, Morocco, 2006).

praised her country's cooperation with Morocco as a success story because of the decline in the number of boats carrying undocumented migrants. She asserted, "The decline is not due to the economic crisis affecting Spain but is the result of stronger control operations off the African coast to prevent boats from heading toward the Canary Islands."[77] Such efforts have corresponded to the stepped-up participation of the Moroccan military.[78]

MOROCCO'S PARTICIPATION IN NATO
AND SECURITY FRAMEWORKS

In terms of the Moroccan-NATO relationship, as mentioned in chapter 3, a June 2008 Exchange of Letters formalized cooperation already in place to enhance "interoperability" between NATO and Moroccan forces. The ostensible reason was terrorism, and navigation of the Strait of Gibraltar is obviously a central concern. But interdiction of illegal boats and other illegal activities are also targeted. This interoperability is conducted under the rubric of Operation Active Endeavor. The relationship was further solidified in October 22, 2009, with the formalization of the agreement in Naples. The Agence Maghreb Arabe Presse (MAP) often hails NATO officials' visits to Rabat, with reports of Morocco's "cooperation on security," "regional solutions," and role as a key "strategic partner" for problems that include WMD, drug interdiction, energy supplies, climate change, and terrorism.

Beyond Morocco's security and military efforts, also important is close multilateral collaboration. In addition to being involved in the Aeneas programs examined in chapter 3, Morocco has been an assiduous participant in MIEUX (Migration: EU Expertise), a joint venture of the EC and the International Centre for Migration Policy Development in Vienna that is underwritten by the European Commission. The Dialogue for Mediterranean Transit Migration (MTM), which involves Arab Partner States (APS) and Europe Partner States (EPS), began with the Alexandria Consultations in 2003. It has continued semiannually ever since, with interior officials meeting in regional capitals to identify migration routes, examine the volume and composition of migration flows, and improve efforts to

77. Quoted in MAP, June 1, 2009. Available at www.map.ma.
78. A fascinating video of the Moroccan National Guard "in action" is available at www.youtube.com/watch?v=PEImrPB08P8.

manage migration.[79] The MTM Dialogue is a multilateral effort, no doubt, but it is rooted in a security logic and aims to enhance the management capacities of states, not to reduce causes of migration or improve treatment of migrants.[80]

CLIMATE-INDUCED MIGRATION AS A NEW RATIONALE

In official documents and discourse, the notion of environmental factors as contributing to migration flows from the Sahel and sub-Saharan Africa did not appear until the end of the last decade. This is not surprising; as detailed in chapter 3 a similar evolution occurred in the North Atlantic context. The Royal Palace and its *Makhzen* is the central actor in the framing of national discourse, and language concerning CIM only recently emerged. In addition to the speeches quoted at the outset of this chapter, on December 8, 2009, Prime Minister Abbas El Fassi read a speech of Mohammed VI's to the second EU-Africa Summit in Lisbon. Not surprisingly, it made extensive mention of the challenges posed by migration from Africa. Yet, without offering evidence of its scope or directionality, he emphasized the environmental and climatic causes of migration:

> Environmental problems are among the main challenges confronting Africa. Growing forest depletion, creeping desertification, soil degradation, drought, poor access to water and climate change compound poverty and threaten population stability. This Summit should therefore enable us to give new momentum to our partnership, commensurate with the challenges of sustainable development and climate change. Just as important is the need to enhance African capabilities to control these phenomena and restrict their harmful effects. Dedicated action is needed because of the interdependent nature of environmental

79. For example, see meeting reports at International Centre for Migration Policy Development and Ministère de l'immigration, de l'identité et du developpement solidaire of the République Francaise, *Dialogue on Mediterranean Transit Migration and i-Map Expert Meeting, Paris, 15–16 December* (Vienna, Austria: ICMPD, 2008); and International Centre for Migration Policy Development and Syrian Arab Republic Ministry of Interior, *Dialogue on Mediterranean Transit Migration and i-MAP Expert Meeting, Damascus, Syria, 30 June–1 July* (Vienna, Austria: ICMPD, 2009).

80. International Centre for Migration Policy Development, Europol and Frontex, *Arab and European Partner States Working Document on the Join Management of Mixed Migration Flows* (Luxembourg: Office for Official Publications of the European Communities, 2007).

problems and development requirements. Such an approach can contribute to greater population stability and better control of migrant movements.[81]

Similarly, the EU-Morocco Summit held in Spain on March 7, 2010, was the first summit between the EU and a Mediterranean partner country since the 2007 Treaty of Lisbon. Spain's Zapatero was keen on hosting the summit in Granada, and Morocco's El Fassi met with Herman Van Rompuy, president of the European Council, and José Manuel Durão Barroso, president of the European Commission. The joint statement emphasized at considerable length Morocco's strategic position vis-à-vis Europe and Africa with declarations such as, "The African continent remains at the centre of [the Parties] common concern."[82] It further notes, "[The Parties] stress the active part played by Morocco in the Africa-EU common strategy, in particular on climate change and strengthening cooperation in the field of peace and security." It continues with:

> The precariousness of the situation in the Sahel region and the many associated challenges show the need for increased regional cooperation and an integrated approach in the fields of security and development. Morocco and the EU consider that the Sahel is a priority zone for action to combat terrorism and radicalisation. Cooperation must be developed between the EU, Morocco and the other countries in the Sahel-Saharan region to take effective action against the threats to security which hang over the region.

The next paragraph adds a resounding point regarding immigration:

> With regard to migration, Morocco and the EU agree to reinforce the mechanisms for cooperation between the countries of origin, transit and destination by pursuing the dialogue between the two Parties and supporting the process of reinforcing the capacity of the parties concerned to combat illegal immigration, promote legal migration, optimise the contribution of migrants to development and deal with the underlying causes of migration. Such a comprehensive and balanced approach to migration issues, also involving cooperation on the return and readmission of illegal immigrants, must constitute a fundamental element of the EU-Morocco partnership.[83]

81. King Mohammed VI, "Speech to the 2nd EU-Africa Summit hosted by Lisbon," Lisbon, Portugal, December 8, 2009.
82. Council of the European Union, "Joint Statement European Union-Morocco Summit, Granada, 7 March 2010," 7220/10 (Presse 54), Brussels: Europa, available at www.consilium.europa.edu/newsroom.
83. Ibid.

Such a declaration, with Moroccan participation—and agreed to upon Spanish soil—simply would have been unthinkable 10 years ago.

CONCLUSION

Throughout the last decade, Morocco increasingly made the interdiction of illegal immigration from the south a strategic priority. It did so to curry favor with European officials, establish its claims on the Western Sahara, and strengthen the status of its own émigrés abroad. Initially, most analyses of the immigration pressures into Morocco did not dwell on environmental dynamics. Instead, they centered on (a) the push from low levels of economic development and civil unrest and (b) Morocco's proximity and ostensibly easy access to Europe. This has begun to change, however, mirroring concomitant changes in the North Atlantic discourse.

How this will evolve in the years to come remains unclear. For Moroccan officials, the CIM card is a mixed bag. To invoke it raises an obvious point: if pressure from the south is increasingly caused by climate change, then Moroccans must be vulnerable to climate change, too. In fact, as chapter 2 demonstrated, the Maghreb and the Mediterranean Basin are more vulnerable than much of the Sahel and equatorial Africa.

Meanwhile, the politics of migration has contributed to a transformation in the Moroccan state. The twin dynamics of (a) Moroccan emigration to Spain and (b) Morocco's role as a point of transit for migrants from the south seeking access to the Canary Islands, the Spanish mainland, and the Spanish enclaves of Ceuta and Melilla have prompted a change in the character of governance and diplomatic posture. What has emerged is a situation quite different from that of the '60s and '70s. Migration dynamics have affected Moroccan sovereignty—not diminished it or enhanced it per se (as if it were readily measurable), but significantly altered it.

A Weberian notion of the state presumes not only a relatively permanent population but also a sense of sovereignty and an ability to control the territory. This implies that the state can determine who is a citizen and can control borders so that noncitizens are kept out. Constructivist theories, however, have illustrated the ways in which states shape their sovereignty in the context of mutually constitutive relationships, especially with neighbors.[84]

84. Alexander Wendt, "Anarchy Is What States Make of It," *International Organization* 42: 2 (1992), 391–425; and Martha Finnemore, "Constructing Norms of Humanitarian Intervention," in *The Culture of National Security: Norms and Identity in World Politics*, ed. Peter Katzenstein (New York: Columbia University Press, 1996), 153–185.

Territorial agreements, border arrangements, and relative degrees of empirical sovereign control are fashioned in the historical interplay of relations between states. The U.S.-Mexican border in this decade is different from last decade's, and more different still from that of the '50s. The very same could be said about the Moroccan-Algerian and Moroccan-Spanish borders. Morocco's "border" with the Western Sahara is even more fraught. In the Maghrebi context, in particular, the colonial legacy and the subsequent dynamism of European regional integration have profoundly affected borders as well as state sovereignty. Prior to Spain's accession to the EEC in 1986, the Spanish-Moroccan border existed, yes, but it was hardly as reified as it is now. By the late '00s, however, the Spanish-Moroccan border had become a tightly patrolled, militarized entity that separates southern Morocco and the Western Sahara from the Canaries, parallels the Strait of Gibraltar, and surrounds Ceuta and Melilla.

Traditional sovereignty—a conceptualization of the state as autonomous, unitary, and separate—is often consistent with the assumptions of realist international relations theories. The state is distinct and individualistic, a man alone. There is also an assumption of immutability and stability. What we see in the case of a transit state, though, is a relational sovereignty born of exchange and interaction. Population issues are especially vexing in the context of a transit state. It was fashionable in the '90s in the works of Sassen and other critics of globalization to argue that population movements undermine a state's sovereignty.[85] But the argument here is that while the movement of populations into and out of a transit state complicates diplomacy, it may in fact *enhance* sovereignty and a state's status in relation to its neighbors and counterparts. This is not to suggest that a transit state's sovereignty is on par with that of other countries. All sovereignty is not created equal. A transit state is in an asymmetrical, subordinate space vis-à-vis North Atlantic neighbors.

Immigration has had many implications for Moroccan politics. First, with respect to human rights, Morocco has adopted the European discourse of stopping illegal migration. For Valluy, Morocco has joined the "ideological armada" that justifies the building of walls to block population movements.[86] At the same time, Morocco challenges several elements of these policies and turns them in a different direction. For example, it refuses to readmit large numbers of irregular migrants from third countries; it will

---

85. Saskia Sassen, *Globalization and Its Discontents: Essays on the New Mobility of People and Money* (New York: Free Press, 1998).
86. Valluy, "Aux marches de l'Europe: des 'pay-camps,'" 341.

not accept Malians or Senegalese whom Spain wishes to send back across the Strait of Gibraltar or the passage between Tarfaya or Laâyoune and Fuerteventura in the Canaries. On a related note, in contrast to Libya and Tunisia, Morocco has rejected the establishment of processing centers for immigrant and asylum claimants on its territory. This comes in the context of Morocco's ongoing effort to fend off international criticism of its human rights record.[87] Since the '70s and the height of the "years of lead," the country has sought to improve its human rights standing. In some aspects this appears genuine, in others more symbolic. Contending with international norms on migration is part of an intricate norms dance. Morocco's domestic human rights organizations are vocal and, while constrained, can be effective. In the context of the transit migrant issue, Morocco agreed in 2007 to give the UNHCR full-fledged representation in Rabat.

Not to be underestimated, too, are the implications for relations with sub-Saharan countries. Morocco has worked intently on African policies— not on the order of Qaddafi's efforts over recent decades, but certainly in terms of trade and diplomacy. Rabat is especially keen on trying to woo support for Western Sahara diplomacy. The expulsion of Africans to their country of origin and incidents such as the 2005 Ceuta/Melilla tragedy complicate those efforts. If Morocco is seen as sovereign over the passageway between the Western Sahara and the Canaries, then its claims become more ensconced. European diplomats have spoken about the "hypocrisy" associated with criticizing Morocco for its assertion of sovereignty over the Western Sahara yet working with Rabat on access to fisheries off the Western Sahara coast.[88]

At bottom, the transit state is in a Janus-faced situation, looking both inward and outward. To enhance its diplomatic credibility, it selectively embraces the rhetoric of fighting illegal migration. In the best of all possible worlds, it would likely prefer not to have transit migration. It is incorrect, however, to suggest that a transit state is *forced* to comply with the interests of advanced-industrialized countries. Instead, transit migration

87. Sieglinde Gränzer, "Changing Human Rights Discourse: Transnational Advocacy Networks in Tunisia and Morocco," in *The Power of Human Rights: International Norms and Domestic Change*, eds. Thomas Risse, Stephen C. Ropp, and Kathryn Sikkink (New York: Cambridge University Press, 1999), 109–133; Slyomovics, *The Performance of Human Rights in Morocco*; and Susan Waltz, "The Politics of Human Rights in the Maghreb," in *Islam, Democracy, and the State in North Africa*, ed. John Entelis (Bloomington: Indiana University Press, 1997), 75–92.

88. Gregory White, "Too Many Boats, Not Enough Fish: The Political Economy of Morocco's 1995 Fishing Accord with the European Union," *Journal of Developing Areas* 31: 3 (1997), 313–336.

is part of a package of sovereignty and diplomatic actions. From the perspective of North Atlantic countries—the ultimate destination for migrants—the transit state is a buffer that presents an opportunity to thicken the border. For the transit state, however, the migration issue might facilitate closer collaboration and ties with advanced-industrialized neighbors. Morocco (or Tunisia, Libya, or Turkey) will not soon be able to reborder itself as part of Europe, as Spain did during its transition. Nonetheless, getting tough on transit migration will continue to facilitate the interests of the state. And mimicking the North Atlantic discourse of CIM can be expected to deepen given its incisiveness. It is reasonable to expect this to continue, unless the dominant discourse is challenged.

CHAPTER 5

Pulling Back the Curtain on the Security Oz

Multilateral Governance and Genuine Sustainability

in a Warming World

All of this cannot leave Europeans indifferent. The Sahara's status as a transit-route for cocaine coming into Europe makes the region a major security issue for European Union countries. And it was groups with roots in north Africa that carried out the metro-bombings in Paris and the train-bombings in Madrid. Until now, Europe's main effort in the region has been to stop African boat-people from entering the EU, making the Sahara into the first line of defence of "fortress Europe". But technical means are not enough. The Europeans need to focus much more attention on what really happens in the Sahara.
—Stephen Ellis[1]

Thinking in security terms is relatively easy. Identify a phenomenon as an imminent threat, prepare to meet it, and reassure yourself (and perhaps others ostensibly counting on you) that the threat has been addressed. If the preparation is feared to be inadequate—as it often is—express alarm and call for further action and ongoing vigilance. The problem, however, is that while the threat may be plausible, its likelihood may not be fully understood. Security discourse can burgeon, too, so that the pursuit of security becomes a fixation rather than a means for meeting genuine threats.

1. Stephen Ellis, "The Sahara's New Cargo: Drugs and Radicalism," *Open Democracy*, April 14, 2010, available at www.opendemocracy.net.

In the performative space of the security theater, one can find unlikely intellectual and policy bedfellows. These liaisons can take a variety of forms. First, for example, a thinker may wholly avoid calling for a harsh security response yet offer an analysis that points to a phenomenon as an emergent concern. In turn, a security-minded analyst can use that analysis to support his or her position. A second form of alliance can emerge when an analyst makes a calculated decision to enlist security intellectuals as allies if securitizing an issue will get policy makers' attention: "You might not care about X, but you should because it is a threat to security." A third form can appear when a cynical opportunist emphasizes the threat and calls for enhanced precautions if it serves his or her own agenda and interests. Finally, the audience can play a part, too, by embracing a security discourse to assuage its worries, even when skeptical about the fundamental causes of the threat.

This is the challenge confronting analyses of climate-induced migration. In recent decades, CIM has clearly become a topic of concern from a wide array of disciplinary and normative perspectives. The argument here is that climate change is occurring and that humans are contributing to it. It is also clear that the age-old human disposition for migration has been and will continue to be profoundly influenced by demographic dynamics, environmental stress, and anthropogenic climate change. In some instances, CIM can lead to conflict, too. Nevertheless, to treat CIM as an emergent security threat to North Atlantic interests is misguided, maintaining the very aspects of the international system that contribute to the phenomenon.[2] The logical prescription is to deepen border security, enhance support for transit states, heighten cooperation between state security and intelligence apparatuses, and treat migrants as desperate lawbreakers. This plays into the agendas and interests of various players, including law enforcement officials and traffickers.

Securitizing CIM not only fails to solve the problem, but is also imprudent because it enhances security against a non-threat. The bulk of CIM has been and will likely be within countries and regions of the world's tropical regions, not directed toward North Atlantic borders. Treating it as a threat sets in motion a counterproductive, spiraling security dilemma and saps energies away from scientific analyses of the phenomenon, from the development of integrative policy solutions devoted to adaptation to

2. Paul J. Smith, "Climate Change, Mass Migration and the Military Response," *Orbis* 51: 4 (Fall 2007), 617–633. See also Joshua W. Busby, "Who Cares about the Weather? Climate Change and U.S. National Security," *Security Studies* 17 (2008), 468–504.

climate change already underway, and from efforts to mitigate GHG emissions. North Atlantic policy makers and electorates may be reassured by border security efforts, which may be cast as a low-cost precaution against climate change's impact. Yet they do not address the underlying issues that prompt the vulnerability.

Moreover, the securitization of CIM contributes to ongoing efforts to reify borders. Borders are spatial and temporal constructs in the Westphalian state system. The Lord Curzon of Kedleston wrote in 1907, "Frontiers are indeed the razor's edge on which hang suspended the modern issues of war and peace, of life or death of nations."[3] They may be a razor's edge in terms of their starkness on a map, but Lord Curzon might better have written "razor wire" or "electrified fence." As argued in chapter 1, it is a mistake to think of borders as thin, immutable lines. In fact, they are dynamic and with varying degrees of thickness depending on the historical and political context. Even the same border can vary in breadth and strength. The U.S.-Mexican border at Tijuana (with California) is very unlike the same international border at Coahuila (with Texas). A thick border is felt miles and miles before it is reached, whereas a thin border might be crossed unnoticed. The bottom line, as geographers and IR theorists have demonstrated, is that powerful players can selectively invoke and reaffirm borders to further political goals.[4]

The constructed character of borders makes them ethical sites as well, with ontological dimensions that are the product of human agency and choices.[5] Reaffirming a border against outsiders is a political act. If outsiders seek access because of injustice—they lack food and shelter, while those on the inside are comfortable—then a fence becomes an especially glaring display of power. And if the people on the inside are consuming inordinate amounts of energy, enjoying luxury items, and leaving behind large amounts of waste and pollution, then the political act becomes a question of justice as well. If those people are aware that others are deprived of basic needs, they might feel guilty. Then again, they might not. To reduce it to the local level, what do people eating on the terrace of

3. Quoted in John Williams, *The Ethics of Territorial Borders: Drawing Lines in the Shifting Sands* (London: Palgrave, 2006), 22.

4. John Agnew, *Geopolitics: Re-Visioning World Politics*, 2nd ed. (New York: Routledge, 2003); David Newman, "Boundaries," in *A Companion to Political Geography*, eds. John Agnew, Katharyne Mitchell, and Gerard Toal (Malden, MA: Blackwell, 2003); and R. B. J. Walker, "State Sovereignty and the Articulation of Political Space/Time," *Millennium* 20: 3 (1991), 445–449.

5. Williams, *The Ethics of Territorial Borders: Drawing Lines in the Shifting Sands*, 9. See also William Walters, "Mapping Schengenland: Denaturalizing the Border," *Environment and Planning D: Society and Space* 20 (2002), 561–580.

a stylish sushi restaurant think when they see a homeless person watching them? What kind of psychological and moral rationalizations does one go through consciously or unconsciously to avoid a sense of obligation? Responsibility? Guilt?

From a Rawlsian perspective, one would likely agree to a political system in which everyone enjoys a degree of fairness.[6] For Rawls, situating the individual in an original position behind a "veil of ignorance"—where the individual does not know where he or she would be positioned in a subsequent political order—would lead a reasonable individual to choose a system based on relative equality. As Vanderheiden argues:

> Behind the veil of ignorance, we may still attempt to choose principles that serve our self-interest, but we are deprived of the relevant knowledge about our particular interests that might allow us to bias the rules in our favor. Behind the veil of ignorance, then, we can chose principles that are fair for all, since each person in effect takes on the position of all. In such a position, what principles of justice would a rational person, aiming to maximize his or her share of society's primary goods, choose? . . . Rawls assumes that persons are risk averse and so would choose principles of justice that maximize the primary goods of the least advantaged—because they may find themselves among that group, for all they know—rather than gambling on the probability that they might wind up among society's more advantaged.[7]

In other words, no one in his or her right mind would choose a system in which he or she would likely end up as the homeless person watching others enjoy endangered *maguro*.

At the level of the international state system, a Rawlsian approach is perhaps less satisfying, at least in terms of its ability to point up the ethical questions at hand. For Rawls, taking the question of the social and moral contract to the international level was less about stark inequalities between states or about people seeking access across borders than about tolerating differences between political orders.[8] Rawls offers very little in *The Law of Peoples* about how we might think about, say, the inequality in the standard of living between Africa and Europe.[9] In terms of international relations theory, to put it conveniently, Rawls accepted the realist vision of unitary,

6. John Rawls, *A Theory of Justice* (Cambridge, MA: Harvard University Press, 1971).
7. Steve Vanderheiden, *Atmospheric Justice: A Political Theory of Climate Change* (New York: Oxford University Press, 2008), 52.
8. John Rawls, *The Law of Peoples* (Cambridge, MA: Harvard University Press, 1999).
9. Vanderheiden, *Atmospheric Justice: A Political Theory of Climate Change*, 103.

sovereign states, with some more republican than others. Moreover, at an international level, one's enjoyment of privilege is not thrown into stark relief by visible suffering. Thick borders are often miles away.

In a warming world, then, how are we to think ethically about borders and borderlands? If a thickened border is used to exclude people who need relief from climate change, then it becomes an ethical site. If the people on the privileged side of the border harden it to cement their privilege, especially when they are responsible for the energy consumption and GHG emissions that cause the problem, then it becomes callous and unconscionable. In addition, here is the irony and the challenging part: the thickening of the border is increasingly being justified by a threat that is not imminent. Migrants driven by climate change are not coming to the border nor to transit states in enormous numbers, nor are they reasonably expected to do so in the decades to come. Significant numbers of people are living in vulnerable, precarious states and are being affected by climate change, yes; they will have to continue to adapt to its challenges. But they will likely do so in regional contexts.

To push the sushi restaurant metaphor (perhaps too far), people are homeless, yes, and they are adversely affected by climate change. Yet they are not moving toward the restaurant. They are not even near the restaurant. They are many miles away. And they are not likely coming, at least not in appreciable numbers or soon. Why then invoke the threat and build a security fence around the restaurant? Does it enhance the enjoyment of a meal knowing that others are deprived? What political and ideological exercise is at work in cementing the privilege of the diners? From a purely self-interested perspective—or even a "self-interest rightly understood" à la de Tocqueville—would it not be better to work on eliminating the causes of the reasons that people might move toward the restaurant in the first place? From an ethical perspective, might it be better to figure out a more just system of food distribution?

How, then, might CIM be "desecuritized"? The important initial step is to pull back the curtain on the security Oz and point to the ways in which the discourse is articulated by real people. This is not to suggest that people have the same motives as the Wizard of Oz, operating controls while hiding behind a curtain. But it does mean being watchful for the myriad ways in which migration is alarmingly invoked as an outgrowth of climate change. Further, by examining empirical reality, it becomes apparent that it does not coincide with the dominant threat-defense discourse.

Moreover, what else could be done to meet the real problems posed by CIM? How could an alternative response be broached, encouraged, and implemented? In this concluding chapter, attention is devoted to two

dominant approaches: (a) global governance and (b) development and climate. Both are more fruitful than a securitized response to the challenges posed by CIM. Global governance emphasizes the need for an institutional, cooperative response on the part of the international community. Development and climate approaches emphasize the need to fashion far-seeing policy solutions at local, national, and global levels of analysis—policies devoted to the emergence of just and ecologically sustainable social, political, and economic institutions. As demonstrated below, strains of the global governance approach can coincide with a security framework, but other elements of it can overlap most fruitfully with development and climate approaches.

GLOBAL GOVERNANCE

An approach that emphasizes the development of international institutions, rooted in the liberal tradition of international relations, has emerged as an effort to establish a regime on behalf of climate refugees. In such an approach, the anarchy of the international system requires an attention to rules, institutions, and regimes. Although global governance typically shies away from interrogating borders and their ethical dimension, its various strains offer important insight into the debates about CIM.

This approach has emerged on both sides of the Atlantic, but it seems to have appeared first in Europe. The Global Governance Project, in particular, is a consortium of European universities and research institutes that emerged in 2001. It includes institutions such as the Vrije Universiteit Amsterdam, Science Po Bordeaux, the University of Bremen, the Freie Universität Berlin, and the London School of Economics. Other related organizations and institutions include the Earth System Governance Project, the International Human Dimensions Programme on Global Environmental Change (IHDP), the United Nations University, and the German Development Institute. This epistemic community emphasizes the importance of non-state actors, emergent norms in international society, and the bridging of local and global issues in efforts to move toward the prevention and mitigation of, and adaptation to, climate change.[10] By way of defining "governance," Haas suggests:

> Governance thus consists of formal institutions designed to obtain collective goals generated from intersubjective beliefs and aspirations. International

10. Frank Biermann, "'Earth System Governance' as a Crosscutting Theme of Global Change Research," *Global Environmental Change* 17 (2007), 326–337.

environmental governance is a process, but one which is principally impelled by changes in formal and informal institutions.[11]

Advocates of this perspective see *governance* as deriving from the Greek word for navigation and invoke metaphors such as "charting a new course."[12] Most intriguing, perhaps, is the call for interinstitutional cooperation. Many of the world's global problems receive crucial attention from single institutions and single regimes.[13] Yet an implicit normative goal is to examine the deepening of "regime interlinkages" and "institutional interplay" into a broader architecture.[14] Additionally, the state is viewed as needing to be adaptive, moving beyond the complex interdependence of the '70s to a deeper capacity for cooperation with other states and institutions.[15]

This is consistent with the "fragmented regime complex" that has developed since the 1992 UNCED in Rio. Between 1992 and the COP15 of 2009, it became apparent that an international organization on the order of the World Trade Organization (WTO)—which is relatively comprehensive in its coverage of trade issues and politics—was not going to emerge for climate change. The United Nations Environment Program (UNEP), for its part, seems to be ill-suited to play a fully integrative role.[16] But as Keohane and Victor have pointed out, there is a vast continuum between an anarchic

11. Peter Haas, "Social Constructivism and the Evolution of Multilateral Environmental Governance," in *Globalization and Governance*, eds. Aseem Prakash and Jeffrey A. Hart (New York: Routledge, 1999), 104. The literature on global environmental governance is voluminous. See inter alia Laura Campbell et al., *Global Climate Governance: Inter-Linkages between the Kyoto Protocol and Other Multilateral Regimes* (Tokyo, Japan: United Nations University Institute of Advanced Studies and Global Environment Information Centre, 1999); Margaret P. Karns and Karen A. Mingst, *International Organizations: The Politics and Processes of Global Governance*, 2nd ed. (Boulder, CO: Lynne Rienner, 2009); and Jennifer Clapp and Peter Dauvergne, *Paths to a Green World: The Political Economy of the Global Environment* (Cambridge, MA: MIT Press, 2005).

12. Frank Biermann, Philipp Pattberg, and Fariborz Zelli, eds., "Global Climate Governance Beyond 2012: An Introduction," in *Global Climate Governance Beyond 2012: Architecture, Agency and Adaptation* (New York: Cambridge University Press, 2010), 1–12.

13. Oran Young, *International Governance: Protecting the Environment in a Stateless Society* (Ithaca, CA: Cornell University Press, 1994); and Peter Haas, Robert Keohane and Marc Levy, eds., *Institutions for the Earth: Sources of Effective International Environmental Protection* (Cambridge, MA: MIT Press, 1993).

14. Frank Biermann and Ingrid Boas, "Protecting Climate Refugees: The Case for a Global Protocol," *Environment* 50: 6 (November/December 2008), 8–16.

15. Frank Biermann and Klaus Dingwerth, "Global Environmental Change and the Nation State," *Global Environmental Politics* 4: 1 (2004), 1–23.

16. Maria Ivanova, "Moving Forward by Looking Back: Learning from UNEP's History," *Green Planet Blues*, ed. Ken Conca and Geoffrey Dabelko, 4th ed. (Boulder, CO: Westview, 2010).

chaos, at one end of the spectrum, to a comprehensive body à la the WTO at the other end. The emerging fragmented regime complex for climate change can, in their view, be more flexible and innovative than a comprehensive regime.[17] There is no common treaty, as many arrangements are informal with different memberships. Quite literally, there is a "dis-integration" of institutional arrangements. The array of instruments in this regime is vast: UN legal regimes such as Kyoto; funding mechanisms such as the Global Environment Facility (GEF); expert assessments such as the IPCC; specialized UN agencies such as UNEP and the Food and Agricultural Organization (FAO); bilateral initiatives between countries; clubs such as the G20; multilateral development banks such as the World Bank; unilateral actions on the part of individual countries and even cities, states, and provinces; and protocols, most notably the Montreal Protocol on Substances That Deplete the Ozone Layer. Taken together, these may provide a kind of toolbox of international responses to environmental problems and climate change. The question that remains is whether a regime concerning CIM is workable. Where would it fit into the basket of arrangements that Keohane and Victor identify?

Most prominent with respect to CIM, perhaps, is the work of Biermann and Boas, whose analysis is infused with a vision of expanding the institutional and norms-based discourse associated with protecting "environmental refugees."[18] Above all, they embrace the use of the word *refugee*, rather than CIM, because it (a) encompasses both internally displaced peoples (IDPs) and transborder migration and (b) has strong moral connotations that amplify the legitimacy and urgency merited by the phenomenon. Moreover, they argue that intensifying the mandate of the 1951 Geneva Convention Relating to the Status of Refugees is implausible and risks inverting the mission of the UNHCR, already strapped with protecting roughly 10 million refugees. Similarly, anticipating that a burgeoning number of environmental refugees will prompt conflict, they argue nonetheless that asking the UN Security Council to respond to a "threat to peace" seems ill-advised, too. The UN's cumbersome

17. Robert O. Keohane and David G. Victor, *The Regime Complex for Climate Change* (Cambridge, MA: Harvard Kennedy School Project on International Climate Agreements, 2010).

18. Biermann and Boas, "Protecting Climate Refugees: The Case for a Global Protocol"; and Frank Biermann and Ingrid Boas, "Global Adaptation Governance: The Case of Protecting Climate Refugees," in *Global Climate Governance Beyond 2012: Architecture, Agency and Adaptation*, Biermann et al., eds. (New York: Cambridge University Press, 2010), 255–269; and Frank Biermann and Ingrid Boas, "Preparing for a Warmer World: Towards a Global Governance System to Protect Climate Refugees," *Global Environmental Politics* 10: 1 (February 2010), 60–88.

response to climate policy and the fact that the five permanent members are among the world's largest emitters of GHGs make it poorly suited to addressing CIM.[19]

Instead, Biermann and Boas call for a new regime on climate refugees, one that would be devoted to planning the voluntary, permanent resettlement of afflicted populations. Additionally, they argue for a program to work with governments that are accepting populations for resettlement and for an elevation of the issue so that it receives attention as a global responsibility. Finally, they argue for a protocol on the recognition, protection, and resettlement of climate refugees, which would include a list of specified administrative areas in countries that "have been determined in need of relocation due to climate change." This is congruent with the call made by others for institutional cooperation. For example, Bogardi and Warner similarly applaud the coordination of IOM, UNEP, the United Nations University, and the Munich Re Foundation into a Climate Change, Environment and Migration Alliance.[20]

Other authors such as Martin, Zetter, and Acketoft have provided additional analyses that deepen the potential for a careful institutional approach to CIM. Martin notes that National Adaptation Programmes of Action (NAPA)—examined more fully below—have not taken CIM into account to any significant extent. She advocates for a more robust engagement of resettlement programs into NAPAs. And she also notes that no major destination country has a proactive policy in place to resettle people affected by environmental change.[21] Martin also points to the International Organization of Migration (IOM) as potentially playing a crucial role; it has offered support for research into CIM. On the other hand, she avers that other agencies such as (a) the International Labor Organization's International Migration Program, (b) the UN Population Division in the Department of Economic and Social Affairs, (c) the Office of the High Commissioner for Human Rights (OHCHR), and (d) the UN Office for Drugs and Crime (UNODC) have not explored in a systematic fashion the interconnections between climate change and their policy responsibilities.[22]

19. Francesco Sindico, "Climate Change: A Security (Council) Issue?" *Carbon and Climate Law Review* 1 (2007), 26–31.

20. Janos Bogardi and Koko Warner, "Here Comes the Flood," *Nature Reports Climate Change* (December 11, 2008).

21. Susan F. Martin, "Managing Environmentally Induced Migration," in *Migration, Environment and Climate Change: Assessing the Evidence*, eds. Frank Laczko and Christine Aghazarm (Geneva: International Organization for Migration, 2010), 353–384.

22. Susan F. Martin, "Climate Change, Migration, and Governance," *Global Governance* 16 (2010), 397–414.

Zetter, for his part, argues that climate change must be incorporated into treatments of people impelled to move, emphasizing the importance of a rights-based approach as a way of expanding concern not only to migration but also to people who do not move and have to adapt to local conditions. He is, however, skeptical of adding new international protocols to the existing set of arrangements.[23] Acketoft points to the potential for a regime to emerge in a manner similar to the one designed to protect IDPs in the early '90s.[24] Martin and Warner have further spearheaded an impressive effort by the German Marshall Fund to bring environmental and immigration specialists into a study team on CIM. Initial documents published in June 2010 offer incisive analyses of various aspects of CIM.[25] Overall, the thrust of the project is to nurture North Atlantic policy dialogues that "balance domestic interests with the clearly humanitarian implications of climate change induced displacement."[26]

Global governance, broadly conceived, is appealing and marks an improvement over the security discourse. It is motivated by a laudable emphasis on social justice and international cooperation. Some of its most passionate examples, such as Byravan and Rajan's analysis of future "climate exiles" affected by rising sea levels, are driven by an ethical preoccupation of the first order.[27] It is also persuasive, because it accepts the fundamental findings of natural science on the projected impacts of climate change.

Nonetheless, three concerns persist. First, ironically, with the exception of migration scholars such as Martin and Zetter, advocates of the global governance approach regarding CIM do not always engage well the findings on migration. Rather than taking a specific migration system to examine as a case—or engaging the broader literature on migration— global governance approaches often accept as axiomatic that climate change will prompt migration on a massive, cataclysmic scale. Although

23. Roger Zetter, "The Role of Legal and Normative Frameworks for the Protection of Environmentally Displaced People," in *Migration, Environment and Climate Change: Assessing the Evidence*, eds. Frank Laczko and Christine Aghazarm (Geneva: International Organization for Migration, 2010), 385–441. See also Roger Zetter, "Legal and Normative Frameworks," *Forced Migration Review* 31 (2008), 62–63.

24. Tina Acketoft, "Environmentally Induced Migration and Displacement: A 21st Century Challenge" (Strasbourg, France: Council of Europe Committee on Migration, Refugees and Population of the Parliamentary Assembly, 2008). See also Martin, "Climate Change, Migration and Governance."

25. See www.gmfus.org for an archive of papers.

26. Susan F. Martin, "Climate Change and International Migration," available at www.gmfus.org, June 2010.

27. Sujatha Byravan and Sudhir Chella Rajan, "The Ethical Implications of Sea-Level Rise Due to Climate Change," *Ethics and International Affairs* 24: 3 (2010), 239–260.

their work is impressive on many fronts, Biermann and Boas go so far as to craft a fictional country facing sea inundation. This may be valuable heuristically. "Lowtidia" may well dramatize rising seas against vulnerable islands. But the problem is that although refugees have emerged and will continue to emerge from the inundation of SIDS, much of the migration may not be wholesale and cataclysmic. As noted in chapter 2, migratory decisions are made within the context of complicated strategies within households, communities, and broader migration systems. It would indeed be a cataclysm if a Lowtidia were effaced quickly, but the inundations confronting coastal countries and SIDS are anticipated to be more gradual and complicated.

Second, and again ironically, the evidence for the prospect of environmental refugees is often drawn from security-minded sources such as the *Stern Review*, the UK Ministry of Defense, and the German Advisory Council on Global Change. Close examination of the UK Ministry of Defense report that Biermann and Boas cite approvingly, for example, reveals panicky italicized emphases. Oddly, too, it moves back and forth between a future indicative and a future speculative. It argues:

> A combination of resource pressure, climate change and the pursuit of economic advantage *may* stimulate rapid large scale shifts in population. In particular, Sub-Saharan populations *will* be drawn towards the Mediterranean, Europe and the Middle East, while in Southern Asia coastal inundation, environmental pressure on land and acute economic competition *will* affect large populations in Bangladesh and on the East coast of India. Similar effects *may* be felt in the major East Asian archipelagos, while low-lying islands *may* become uninhabitable.[28]

Bauer evinces similar concern about CIM, yet takes as evidence the WBGU's framing of environmental refugees as a security threat.[29]

In the security performance described at the outset of this chapter, the challenge for the global governance line of argument is to be wary of the alarmed assumptions and speculations of security-minded authors and institutions. Global governance prescriptions may differ from security nostrums, but the diagnosis is similar. The irony is that in the late '80s and '90s, analysts concerned about climate change and its impact on migration

---

28. United Kingdom Ministry of Defence's Development, Concepts, and Doctrine Centre, *DCDC Global Strategic Trends Programme: 2007–2036*, 3rd ed. (London: UK Ministry of Defence, 2007).

29. Steffan Bauer, "Land and Water Scarcity as a Driver of Migration and Conflicts?" *Agriculture and Rural Development* 1 (2007), 7–9.

debated whether it was wise to enlist security discourse. Now security officials have effectively become convinced that CIM is a real concern and have
cast it as a threat, and analysts not necessarily disposed toward a securitized response use the security community's characterization as evidence
for their argument. To be fair, again, arguing that a global governance
approach borrows diagnoses from the security community does not mean
that it shares the policy solutions of robust borders or military cooperation
with transit states. Its proclivity for institutional cooperation is profoundly
liberal and much less realist than the security thinkers'. Nonetheless, it
needs to be mindful of the challenges.

A third predicament is that rather than dwelling on efforts to achieve
fully sustainable development—with an emphasis on both mitigation and
adaptation—the global governance approach seems to accept the inevitability of catastrophic climate change. There is more than a hint of a resigned
acceptance of the worst-case scenario. Hundreds of millions of people *will*
be displaced by climate change, they *will* move toward North Atlantic borders, and we *must* do something (at the institutional, governance level) to
prepare for it. In many ways, the global governance perspective may be
unwittingly congruent with an alarmist, security-minded viewpoint. The
worst is going to happen; it is inevitable. So prepare for it. Rather than
averting massive CIM by dealing with mitigation and adaptation and working to understand the complicated character of human migration, global
governance is calling for an institutionalized response to the worst-case
scenario: the need to "resettle millions of people."[30]

Still, global governance's emphasis on rethinking and spurring greater
attention to the impact of climate change is laudable. When joined with
development and climate initiatives, it could be especially potent.

DEVELOPMENT AND CLIMATE

The challenges posed by the empirical realities and future prospects of CIM
can be met by addressing the nexus of development and climate. Here it is
important to stress *genuine* development and the full fruition of human
potential, not a notion of development that assumes a limitless frontier or
that is devoted to consumption measured as rapid GDP growth. The North
Atlantic's enormous inputs of energy and resources and outputs of pollution

---

30. For an apocalyptic vision of climate change and migration, see David Hood's *Fatal
Climate* (London: Phoenix, 2001). One might even say that it is a left-wing variant of
Crichton's *State of Fear*.

and waste have never been sustainable. It was only in the late '60s, with the emergence of a "limits to growth" ethos, followed by the sustainable development paradigm of the '80s, that the notion of limitless growth was fully challenged.[31]

Indeed, in keeping with the emphasis on the challenges for countries and regions in the tropics, and with the emergent heft of Brazil, Russia, India, and China (BRIC), it makes sense to expand the concern here to members of the newly established Group of 20 (G20). *Both* advanced-industrialized and rapidly industrializing G20 countries must pursue mitigation strategies. Granted, their political circumstances and their levels of development are quite heterogeneous. Nevertheless, taking the G20 as a whole makes good sense: it comprises 90 percent of global GDP, 80 percent of global trade, and two-thirds of the world's population.[32] If it does not reform itself collectively to pursue a greener economy, then projections are indeed dire.

Specifically, G20 countries need to sharply mitigate GHG emissions even as they pursue industrial and economic growth. Advocates of free trade and rapid economic growth are often critical of caps on GHG emissions. Pointing to the United States as an inordinate emitter of GHGs is misguided, they say; it stands to reason that the most powerful economy would produce the most GHGs per capita. Yet there are advanced-industrialized economies that are not emitting as much carbon and that are pursuing mitigation. The International Energy Agency argues for decoupling CO_2 emissions from economic growth and points to the disparate per capita emissions as evidence that growth *can* be pursued without high rates of GHG production.[33]

The best way to slow climate change is to diminish the GHGs that are forcing warming. There is no silver bullet for achieving this goal, but there may be "silver buckshot," or multiple answers. Some of the solutions are low-hanging fruit; they could quickly be implemented with proper political leadership. Improving transportation networks, for example, would go far to reduce reliance on automobiles. Other steps are more dramatic. What is essential is pursuing an array of aggressive policies that seek to move in the right direction with the understanding that there are, indeed, "boundaries

31. Donella H. Meadows et al., "The Limits to Growth," in *Green Planet Blues*, eds. Ken Conca and Geoffrey D. Dabelko, 4th ed. (Boulder, CO: Westview Press, 2010), 25–29; and United Nations World Commission on Environment and Development, *Our Common Future* (New York: United Nations, 1987).

32. See www.g20.org.

33. See Fridtjof Unander, *From Oil Crisis to Climate Challenge: Understanding CO₂ Emission Trends in IEA Countries* (Paris: OECD and IEA, 2003).

for a healthy planet" and that a "safe operating space" can be achieved.[34] These include transitioning to energy-efficient, low-carbon economies so as to stabilize the atmospheric concentration of CO_2; curtailing land degradation by limiting urban sprawl and freshwater depletion; reforming industrialized agriculture by reducing fertilizer use, limiting nitrogen and phosphorous pollution, and improving water use; and increasing sequestration of CO_2 emissions in sinks.

As countries continue to pursue rapid growth, especially in the aftermath of the 2008 global fiscal crisis, governments must be especially attentive and avoid returning to old practices. Growth in the twenty-first century simply cannot continue to be pursued with nineteenth-century ideologies. Tragically, efforts to recover from the 2008 crash seem to have jettisoned green growth. The most notorious fuel of all, coal, continues to be consumed in record amounts, not only by China and Europe but also by the United States. Worse are perverse, Orwellian deployments of "clean coal" language and specious claims that the carbon produced by coal can be captured and sequestered.

For the non-G20 countries—found in Africa, the Middle East, southwest Asia, southeast Asia, Oceania, and equatorial and Andean South America—the challenge is not one of mitigating GHG emissions, although that will be an issue if they choose the path of rapid industrialization and high consumption. Rather, it is one of adapting to climate change. This is a serious challenge, yet there is little other recourse.

Many such adaptations have long been made at the local level. People living in harsher climes, including the tropics, have skillfully adapted for millennia. This is not to romanticize them as close to nature and somehow brave and noble. Nevertheless, one constant is people's ability to adapt to change. At times, too, people respond well to external assistance and support. In the Sahelian and sub-Saharan African context, there is ample evidence that people adapt to even the harshest circumstances. For example, during the brutal droughts of the '70s and '80s, many crop and livestock farmers innovated with outside assistance. Many farmers invested in new technologies so that when the rains returned and a new equilibrium was achieved, they were in a better position. As Reij et al. argue (consistent with the findings in chapter 2), contrary to the dominant discourse of the era and to this day, people in the Central Plateau of Burkina Faso did not migrate to northern Africa and Europe during times

34. Johan Rockström et al., "A Safe Operating Space for Humanity," *Nature* 461 (September 24, 2009), 472–475; and Jonathan Foley, "Boundaries for a Healthy Planet," *Scientific American* 302: 4, 2010, 54–57.

of drought.[35] Instead, they moved to urban centers within Burkina or to coastal countries such as the Ivory Coast.[36] Those who remained experimented with improving soil and water conservation (SWC) techniques, and their success led to more funding. The result has been significant increases in crop yields, more trees (even as the vegetation on nontreated areas continues to decline), greater availability of forage for livestock, improved livestock management, and a decrease in outmigration.

This has coincided with the greening of the Sahel. In some respects, the greening has been characterized as a recovery from the droughts of the '70s and '80s, but that assumes, incorrectly, an equilibrium in the past. Rather, there have been multiple equilibrium points; "greenness" changes over time as part of the natural oscillations of climate variability.[37] Further, in recent decades different parts of the Sahel have greened to different degrees, the western portion more than the eastern. And as Hein and De Ridder argue, the interpretation of satellite imagery and the variability of rain-use efficiency (RUE) may bias findings toward an interpretation of greening.[38] So it must be said that the greening is not fully understood— and that the climate remains very harsh.

One possibility with respect to the Sahel concerns tipping points—by which climate change would occur abruptly, not gradually, as is often assumed (see chapter 2). Lenton et al. argue that the Sahel may, ironically, "benefit" from climate change. In their examination of a variety of AOGCMs, the prospects for the Sahel are unclear. In two AOGCMs, for example, the West African monsoon (WAM) collapses, but the result differs. In one, the collapse leads to further drying of the Sahel; in the other, it leads to increased rainfall due to increased inflow of moisture from the ocean (stemming from an approximately 3°C warming of SSTs in the Gulf of Guinea). A third AOGCM predicts no large trend in mean rainfall but a doubling of the number of unusually dry years by the end of the century. They write:

> Increasing atmospheric CO_2 has been predicted to cause future expansion of grasslands into up to 45 percent of the Sahel, at a rate of up to 10 percent of

35. C. Reij et al., "Changing Land Management Practices and Vegetation on the Central Plateau of Burkina Faso (1968–2002)," *Journal of Arid Environments* 63 (2005), 642–569.

36. Toulmin, *Climate Change in Africa* (London: Zed Books, 2010), 120.

37. L. Olsson et al., "A Recent Greening of the Sahel: Trends, Patterns and Potential Causes," *Journal of Arid Environments* 63 (2005), 556–566.

38. Lars Hein and Nico De Ridder, "Desertification in the Sahel: A Reinterpretation," *Global Change Biology* 12 (2006), 751–758. By contrast, see Stephen D. Prince et al., "Desertification in the Sahel: A Reinterpretation of a Reinterpretation," *Global Change Biology* 13 (2007), 1308–1313.

Saharan area per decade. In the Sahel, shrub vegetation may also increase due to increased water use efficiency (stomatal closure) under higher atmospheric CO_2. Such greening of the Sahara/Sahel is a rare example of a beneficial potential tipping point.[39]

If this greening does come to pass, then the notion of enhanced CIM from the region, based on the assumption that drying leads to migration, is thrown into question all the more.

Additionally, valuable insights continue to be generated on the ground by researchers seeking to understand the prospects for development and climate initiatives. For example, Reynolds et al. argue for a Drylands Development Paradigm (DDP) that would integrate ecological issues, social issues, and attention to the dynamism of climate change.[40] They are especially keen on integrating local environmental knowledge (LEK) with policy solutions. Moreover, nonlinear processes need to be recognized, given multiple equilibrium states with divergent thresholds. Additionally, Raynaut makes a superb argument for rejecting adaptation as a passive strategy:

> Adaptation should not be construed as a passive reaction to external factors: it is a process of innovation and creation strongly related to the cultural and organizational characteristics of social systems. In other words, it relates to individuals' position or "room for manouver" in larger social, political and economical spheres; and also to the confrontations they are involved in between social actors.[41]

Population, health, and environment (PHE) approaches also emphasize adaptation. The Environmental Change and Security Program of the Woodrow Wilson Center regularly features PHE analyses on its superb website, The New Security Beat.[42]

Policy makers need to be as adaptive and dynamic as local stakeholders. And both parties—indeed all parties to these concerns—need to better understand that the biophysical environment changes rapidly. Neither the

39. Timothy Lenton et al., "Tipping Elements in the Earth's Climate System," *Proceedings of the National Academy of Sciences* 105: 6 (February 12, 2008): 1786–1793.

40. James F. Reynolds et al., "Global Desertification: Building a Science for Dryland Development," *Science* 316 (May 11, 2007), 847–851.

41. Claude Raynaut, "Societies and Nature in the Sahel: Ecological Diversity and Social Dynamics," *Global Environmental Change* 11 (2001), 9–18.

42. Roger-Mark De Souza, *Focus on Population, Environment, and Security* (Washington, DC: Environmental Change and Security Program of the Woodrow Wilson Center, 2009), available at newsecuritybeat.blogspot.com.

drought of the '70s and '80s nor the recent greening should have been a surprise. As Warren argues:

> The scientific literature now leaves little room for doubt: semi-arid ecosystems are inescapably dynamic, and rainfall is the main driver. The Sahelian environment will never permanently "recover" from the droughts of the 1970s and 1980s, so much as experience a succession of bad and good phases, whose effect can only be aggravated or mitigated by policy, not eliminated. Surprise at the reports of a recovery is a strong indication that this lesson has not been learned.[43]

Science has to be vibrant and work even harder to help local populations meet their basic human needs. Technology needs to be deployed, but local knowledge also needs to be furthered. Weather stations require instrumentalization and appropriate staffing. This is not a huge cost either. As noted in chapter 2, a low-tech weather station that can be uplinked to the Internet is very inexpensive. A group of parents in Leeds, Massachusetts, secured a grant to purchase a weather station for a public primary school for a few hundred dollars. Providing dozens or even hundreds in Africa would hardly strap budgets. Given the size of defense budgets in Sahelian and sub-Saharan Africa, not to mention in Washington, DC, lack of funds is clearly not the problem. Further efforts also need to be made to stabilize agriculture and natural resource management, improve cereal yields, increase tree density, develop soil management, and combat rural poverty.

A practical response—one admirable in its normative orientation—is the Millennium Villages approach funded by the UN Environmental Program and Columbia University's Earth Institute. Outside experts collaborate with local stakeholders and local and national officials to craft site-specific solutions for providing seeds, fertilizers, medicines, and educational infrastructure. Science and technology are used as well, including pharmaceuticals, geographic information systems (GIS), the Internet, and remote sensing. It is similar to past efforts to pursue integrated rural development, but different in its emphasis on local stakeholders, technology, and commitment to the Millennium Development Goals (MDG). The eight MDGs, articulated at the United Nations' Millennium Summit in New York in 2000, all have capacity development and governance improvement dimensions at the local, national, and multilateral levels. In Tiby, Mali, for

---

43. A. Warren, "The Policy Implications of Sahelian Change," *Journal of Arid Environments* 63 (2005), 667.

example, integrated programs have been devised to contend with the poor soils, erratic rainfall, and human and animal population demands of the Segou region. In an effort to increase agricultural yields, specialists have worked with local farmers to improve the use of sustainable fertilizers, seed varieties, and drip irrigation. In addition, boreholes have been drilled, roads built, off-grid electrical networks constructed, classrooms constructed, and health clinics staffed.[44] All this is hardly sufficient, but it is a step in the right direction and can serve as a model.

Further, at the diplomatic and state level there have been positive developments with respect to National Adaptation Programmes of Action (NAPAs). NAPAs are a central component of the UNFCCC process. At UNCED in Rio, the predicament of tropical countries was front and center. By the seventh COP, held in Marrakech in 2001, the UNFCCC had requested that developing countries establish NAPAs tailored to their circumstances. By the end of the '00s, roughly 40 countries had supplied NAPAs.

The challenge with NAPAs is that they be engineered to truly engage the issues confronting African countries, rather than slip into well-worn grooves. As Seck argues, there has been a bit of déjà vu, with a lack of data, a low awareness of climate change issues, and an absence of coordination mechanisms at the regional and local levels, posing constraints for the implementation of development policy.[45] Efforts to fund local initiatives and solutions—and improve the management of programs—have to be redoubled and retripled.

Additionally, NAPAs tend to exhibit an implicit bias against migration. Many NAPAs do not treat migration at all, and proactive adaptation strategies—wherein urban planning might situate communities away from vulnerable coastal areas or other areas susceptible to flooding—are poorly considered.[46] Put differently, an additional challenge for NAPAs concerns national-level governance, as opposed to the international-level discussed above. National governments must deepen their attention to governance by integrating adaptation and the reduction of vulnerability to the impact

44. Earth Institute and United Nations Development Program, *The Millennium Villages Project: Annual Report 2008* (New York: Columbia University, 2008), available at www.millenniumvillages.org.

45. Emmanuel Seck, "National Action Programs for Climate Adaptation: A Déjà Vu or a Real Chance to Build on Past Experiences? A Presentation to the ACP-EU Council of Ministers-Specialist Conference Report on Governance and Combating Desertification" (Brussels, Friedrich Ebert Stiftung, May 23, 2007). See also Toulmin, *Climate Change in Africa*, 28–29.

46. Ronald Skeldon, "Background Paper for Roundtable Session 3.2: Assessing the Relevance and Impact of Climate Change on Migration and Development," Global Forum on Migration & Development, Mexico, November 2010, available at www.gfmd.org.

of climate change. The COP15 in Copenhagen, and the December 2010 COP16 in Cancun, Mexico, made some progress in this regard by broaching and pledging to provide $30 billion in "fast-start" monies for adaptation and mitigation. Much of this funding—as well as an additional $100 million pledged to developing countries by 2020 at Cancun—will go far to improve country-level governance, if it is well implemented. As Raleigh argues convincingly, the primary cause of conflict in Sahelian and sub-Saharan Africa is not necessarily exposure to a precarious, forbidding environment, but rather a vulnerability borne of poor governance from country to country.[47] Improving that governance capacity reduces the political marginalization of key populations, not to mention the potential for migration spurred by climate change.

CONCLUSION

The 2008 global fiscal crisis demonstrated that the international financial system is so intricate and intertwined that unilateral responses by individual countries are inadequate. What was necessary was a multilateral approach. To assume that a country *could* act in an insular fashion is to accept the simplest axioms of realist theories of international relations or mercantilist approaches to the international economy. Unfortunately, no multilateral approach has emerged, and G20 economies have made only small nods to multilateral cooperation.

The global environmental predicament, too, requires a multilateral approach, ideally one devoted to green development and social justice. In the run-up to COP15 in Copenhagen, prominent analysts frequently called for leadership by the United States on climate change.[48] Yet this seemed to fail to take into account the profound interdependencies within the international economy, as if the United States could somehow do something separate from its close ties with the EU, or act independently of Niall Ferguson's neologism "Chimerica." Getting individual North Atlantic governments to agree to a reduction of CO_2 emissions to the amount called for by the IPCC—25–40 percent below 1990 levels by 2020—was unlikely at Copenhagen. Most disappointingly, North Atlantic leaders did not adequately address their electorates and urge them to support policies designed

---

47. Clionadh Raleigh, "Political Marginalization, Climate Change and Conflict in African Sahel States," *International Studies Review* 12 (2010), 69–86.

48. See, for example, Jeffrey Sachs, "US a Laggard in Climate Negotiations," *The Times of India*, October 28, 2009, available at www.timesofindia.indiatimes.com.

to mitigate GHG emissions and help developing countries move toward mitigation and adaptation. Better progress seems to have been made at Cancun in December 2010, primarily because expectations of major breakthroughs were tempered.

Despite the uncertainties of multilateral solutions, important steps are being taken to address the challenges posed by climate change, development, and human migration. The better ones are eschewing a securitized discourse. In addition to the global governance approaches detailed above, another promising, recent opportunity for the evolution of international norms is the Global Forum on Migration and Development (GFMD), an initiative of the United Nations. The GFMD is part of an effort to enhance knowledge, improve data gathering, pursue sustainable development, and avoid securitized discourse. Migration discourse within the research prepared for the November 2010 meeting of the GFMD in Mexico was devoted to viewing migration as a proactive adaptation strategy, and the need to incorporate it into planning at the local, national and international levels.[49] Additionally, there is considerable anticipation for the United Kingdom Government's Office for Science release of an October 2011 project, "Foresight Global Environmental Migration." How such a report treats the security dimension of CIM itself, or, at a meta-level, whether the report is read by the international community as providing evidence for a security-minded response will be, to say the least, very important.

Sebastian Junger's 1997 book *The Perfect Storm*, while gripping, introduced a now overused metaphor into common parlance.[50] A "perfect storm" suggests a rare congruence of unfortunate, and unavoidable, circumstances. The phrase aptly described the 1991 nor'easter—a ferocious meeting of low-pressure systems in the Atlantic—that claimed the swordfishing boat *Andrea Gail*. But too often one hears a leader invoking a perfect storm as an excuse for a university's budget crisis, a sports team's poor performance, or a company's dismal earnings report. "We were caught off guard by a combination of unforeseen events, such as. . . ." The point of Junger's book, however, was that the captain of the *Andrea Gail* did not heed ample warning signals. Driven by hubris and the economic imperatives of longline fishing, he ignored the weather reports and sailed into the heart of the storm. Similarly, when leaders ignore a variety of gauges tipping into the red zone, and still press on, their perfect-storm excuses should not be accepted.

49. Skeldon, "Background Paper for Roundtable Session 3.2: Assessing the Relevance and Impact of Climate Change on Migration and Development."
50. Sebastian Junger, *The Perfect Storm* (New York: Norton, 1997).

It is common in discussions about climate change to speak of "our children" and "our grandchildren," yet we already live in a different world than our parents and grandparents. Comparing what is known now to what was anticipated at the 1992 UNCED in Rio, or in even earlier expectations of climate change, it is apparent that anthropogenic climate change is well upon us and worsening. The Gravity Recovery and Climate Experiment (GRACE) is a pair of German-U.S. satellites that orbit the Earth in tandem, 300 miles above its surface. Through constant communication, the satellites are able to measure distances between themselves and the surface below, and thereby calculate the differential impact of gravity. The small discrepancies in gravity, in turn, facilitate calculations of landmasses. In the case of the Greenland and Antarctic ice sheets, GRACE affirms that there has been a precipitous loss of ice in recent years, one far worse than predicted.[51]

The political landscape hardly inspires an optimistic assessment of the prospects for dealing with climate change, not to mention climate-induced migration, but we have had ample warning. Any claim of an unlikely "perfect storm" is now unacceptable.

So we know that climate change is upon the Earth. How are we to conceptualize and frame its implications? How do we mitigate it? How do we adapt to its effects? What is clear is that while a security framework is politically successful, it takes intellectual, political, and financial capital away from more fruitful and just policy measures. Greater attention needs to be paid to crafting development solutions that have environmental implications at their core. Policies that address both mitigation and adaptation need to be emphasized at every turn. Local-level governance in countries more vulnerable to climate change needs to be strengthened. Finally, international multilateral solutions must be robust. By far, the least constructive effort is to emphasize security and militarized borders, which reinforces the unilateral aspects of the international system and undermines sensible policies.[52]

Alas, the treatment of CIM as a security concern continues unabated.[53] Both the 2010 U.S. Quadrenniel Defense Review (QDR) and a concurrent

51. Alexandra Witze, "Losing Greenland," *Nature* 452 (April 17, 2008), 798–802.

52. Jon Barnett, "The Prize of Peace (Is Eternal Vigilance): A Cautionary Editorial Essay on Climate Geopolitics," *Climatic Change* 96 (2009), 1–6.

53. For provocative treatments see Cleo Paskal, "How Climate Change Is Pushing the Boundaries of Security and Foreign Policy," 07/01, Energy, Environment and Development Programme, Royal Institute of International Affairs, Chatham House, London, June 2007; Cleo Paskal, *Global Warring: How Environmental, Economic and Political Crises Will Redraw the World Map* (London: Palgrave Macmillan, 2010); and Gwynne Dyer, *Climate Wars* (London: Oneworld, 2010).

"Green Paper" in the United Kingdom identified climate change as a security issue. In an essay, UK Rear Admiral Neil Morisetti and U.S. Deputy Assistant Secretary of Defense for Strategy Amanda Dory praised the documents and argued:

> Climate change–induced water and food scarcity could spur changes in migration patterns in areas where tensions already run high. With 600 million people living less than 35 feet above sea level, rising waters could cause massive displacement of populations, and could devastate crops and property. We have both concluded that our militaries have a role to play in fostering efforts to assess, adapt to and mitigate the effects of climate change.[54]

North Atlantic policy makers—and analysts, too—need to do better. Casting CIM as a security matter does not attenuate the factors that cause climate change, does not facilitate a just preparation for its effects, and does not help people who have to adapt to it. It merely gives an anxious audience the illusion of security.

EPILOGUE

2011: Contagion, Refugees, and Instability to Europe's South

On January 14, 2011, Tunisia's Jasmine Revolution led to the overthrow of Zine el Abidine Ben Ali, who had ruled Tunisia for 23 years. Ben Ali's removal soon sparked protests throughout the Middle East and North Africa (MENA). On February 11, after 29 years in power, Egypt's Hosni Mubarak also stepped down. And by the end of February, Libya had descended into a bloody, tragic civil war. Countries within and beyond the MENA region experienced sharp upheaval as people challenged authoritarianism, protested rising food and energy prices, and sought a better life.[55]

There is no doubt that the events of 2011 are world historic; historians will analyze their significance for years to come. The skeptical analyst might worry that events may lead to "Ben Ali without Ben Ali" or "Mubarak without Mubarak," wherein the removal of an unpopular leader still leaves in

---

54. Neil Morisetti and Amanda Dory, "The Climate Variable: World Militaries Grapple with New Security Calculus," *Defense News*, March 29, 2010.

55 I examine the unfortunate tendency to cast the events of January 2011 as an "Arab" phenomenon in Gregory White, "It's Not Because They're Arab," *Open Democracy*, February 10, 2011, available at www.opendemocracy.net/gregory-white/it's-not-because-they're-arab.

place fundamental structures. The optimistic take, by contrast, is that they will augur political and economic reforms that will lead to more democratic and more just societies. Time will tell.

Whatever the outcome, this book has argued that climate-induced migration (CIM) has emerged as a security issue and that it is increasingly abetting a logic of control that, unchecked, will likely deepen in the future. The book has further argued that Europe's southern border has been predicated on a policy of "remote migration control"—the extension of the border deeper into Maghrebi transit states. Europe (joined by the United States) has based its support for authoritarian regimes on a narrow logic focused on stability and cooperation in migration control, as well as access to hydrocarbons and repression of political Islamism. This outsourcing of border protection has eschewed a concern for peace, or prosperity, or human rights.

What is especially striking about the Libyan case, for example, is that until the country's collapse in February 2011 the EU and Libya worked closely together on migration control. In June 2010, a Memorandum of Understanding promised to provide EU technical assistance until 2013. Subsequently, in October 2010, a Migration Cooperation Agreement was signed that provided 60 million euros to Libya in order to manage migration flows.[56] On a bilateral level, Italy and Libya had grown closer together, stemming in large part from a Partnership Treaty signed in August 2008 that provided $5 billion compensation for the abuse of Italian colonialism. (Libya is a perfect example of the old adage that the only thing worse than being colonized is being colonized poorly; Italian colonialism was especially depraved.)

Still, it would be wrong to blame Italy alone for supporting Qaddafi's efforts to thwart emigration to Europe. After all, Italy was only complying with EU strictures on border control established by the Dublin Regulation in 2003. In an interview with the BBC on March 1, 2011, Qaddafi expressed genuine bewilderment and anger that his ongoing rapprochement with North Atlantic powers had transformed into calls for his removal. He was asked, "Do you feel a sense of betrayal about that? Did you ever regard [Western powers] as friends?" Qaddafi responded with a vehement, "Of course it's betrayal. They have no morals."[57]

---

56. See European Commission, "European Commission and Libya Agree to a Migration Cooperation Agenda during High Level Visit to Boost EU-Libya Relations," Brussels, October 5, 2010, europa.eu/rapid/pressReleasesAction.do?reference=MEMO/10/472.

57. Excerpts from the March 1, 2011 interview are available at www.bbc.co.uk/news/world-africa-12604102.

As for Morocco, the transit state that provided the case examined closely in chapter 4, much has been made about the fact that it has avoided the turmoil experienced by other countries. Exceptionalist narratives variously focus on Mohammed VI's charisma and legitimacy as a descendant of the Prophet and Commander of the Faithful; the eschewal of single-party government in the early years of independence; the allowance of a moderately robust press, with a few starkly drawn red lines concerning the Western Sahara or the monarchy; the moderate reforms concerning Berber identity and the family code in the '00s; and the king's willingness to explore the human rights abuses under his father, Hassan II. On March 9, 2011, Mohammed VI announced the formation of a commission that would pursue reforms of the constitution by June 2011, including the appointment of the prime minster by the Majlis rather than by the king, as has been the historic practice. These are significant reforms. Whether they forestall the unrest experienced elsewhere remains to be seen.

If Morocco is an exception, it is an exception wrapped in a paradox. On a wide array of social and economic indicators—unemployment, cost of living, rates of literacy and education, poverty, health care, etc.—Morocco is worse off than most of its neighbors. It has long been outpaced by Tunisia, in particular, on such indicators. And on issues of corruption or the crony capitalism that bedevils other countries, the country is second to none. Again, Morocco escaped the widespread instability experienced by its neighbors in the winter of 2011, and it may indeed be "exceptional," but a wager on the future would hardly be a sure bet.

The concern here is that the regional turmoil is not going to cause North Atlantic powers to reevaluate the logic of support for authoritarian structures to their south. To the contrary, alas, it may result in an affirmation that the causes of North Africa's turmoil are endogenous to the region, that the entire region is a zone of insecurity, and that it has an even more fraught and burdened zone to its south—namely, Saharan, Sahelian, and sub-Saharan Africa. What appears to be especially troubling is the way in which the turmoil was framed in early 2011 as a "contagion"—as if the instability is a disease that merits quarantine.

As for the migration piece of these events, in late January 2011, shortly after Tunisia's Jasmine Revolution but before Libya collapsed into full civil war, where was the international media's attention drawn? Not to analyzing the new government, or looking at efforts to reform economic structures, or speculating at the potential impact of Ben Ali's fall on the impressive reforms Tunisia had enacted over 50 years concerning the status of women within society, or reexamining (soul searching?) how North Atlantic policies sustained Ben Ali and his family . . . but to reports of

Tunisians fleeing their country by the boatload and trying to make landfall on the Italian island of Lampedusa. However the pressures that drive mixed migration to Europe are framed, the concern is the same: there is a turbulent zone of insecurity to Europe's south.

Climate-induced migration had receded as an ostensible worry in early 2011. But if and when the regional turmoil calms down, casting CIM as a security risk must be avoided. Catastrophism is a catchy ailment. North Atlantic citizens need to transcend the temptation to become susceptible to panicky assumptions, and policy makers need to avoid a facile security discourse that is politically successful. Instead, we should work to change policies—especially migration and energy policies—that have contributed to the MENA region's turmoil. People in the MENA region and beyond have made it clear that they reject authoritarianism. One also hopes that North Atlantic actors reject the policies that have facilitated authoritarianism in neighboring transit states.

BIBLIOGRAPHY

Acketoft, Tina. "Environmentally Induced Migration and Displacement: A 21st Century Challenge." Strasbourg, France: Council of Europe Committee on Migration, Refugees and Population of the Parliamentary Assembly, 2008.

Adamo, Susana B. "Addressing Environmentally Induced Population Displacements: A Delicate Task." Population-Environment Research Network Cyberseminar on "Environmentally Induced Population Displacements, 2008, available at www.populationenvironmentresearch.org.

Afifi, Tamer. "Niger." In EACH-FOR: Environmental Change and Forced Migration Scenarios D.3.4. Synthesis Report, edited by Andras Vag. Brussels, Belgium: European Commission, 2009, 42–44.

Agamben, Georgio. Homo Sacer: Sovereign Power and Bare Life. Translated by Daniel Heller-Roazen. Stanford, CA: Stanford University Press, 1998.

Agnew, John. Geopolitics: Re-Visioning World Politics. 2nd ed. New York: Routledge, 2003.

Allison, Graham, and Philip Zelikow. Essence of Decision: Explaining the Cuban Missile Crisis. 2nd ed. New York: Longman, 1999.

Allison, Ian, Nathan Bindoff, Robert Bindschadler, Peter Cox, Nathalie de Noblet-Ducoudré, Matthew England, Jane Francis, et al. The Copenhagen Diagnosis: Updating the World on the Latest Climate Science. Sydney, Australia: University of New South Wales Climate Change Research Center (CCRC), 2009.

Andreae, Meinrat O., Chris D. Jones, and Peter M. Cox. "Strong Present-Day Aerosol Cooling Implies a Hot Future." Nature 435: 30 (2005): 1187–1191.

Andreas, Peter. Border Games: Policing the US-Mexico Divide. Ithaca, NY: Cornell University Press, 2000.

———. "Redrawing the Line: Borders and Security in the Twenty-First Century." International Security 28: 2 (2003): 78–111.

Andreas, Peter, and Kelly Greenhill, eds. Sex, Drugs and Body Counts: The Politics of Numbers in Global Crime and Conflict. Ithaca, NY: Cornell University Press, 2010.

Andreas, Peter, and Timothy Snyder, eds. The Wall around the West: State Borders and Immigration Control in North America and Europe. Lanham, MD: Rowman & Littlefield, 2000.

Baldwin-Edwards, Martin. "Between a Rock and a Hard Place: North Africa as a Region of Emigration." Review of African Political Economy 33: 108 (2006): 311–324.

Barnett, Jon. "The Prize of Peace (Is Eternal Vigilance): A Cautionary Editorial Essay on Climate Geopolitics." Climatic Change 96 (2009): 1–6.

———. "Security and Climate Change." Global Environmental Change 13 (2003): 7–17.

Barnett, Jon, and W. Neil Adger. "Climate Change, Human Security and Violent Conflict." *Political Geography* 26 (2007): 639–655.

Barnett, Michael. "The New United Nations Politics of Peace: From Juridical Sovereignty to Empirical Sovereignty." *Global Governance* 1: 1 (1995): 45–61.

Barros, Lucile, Mehdi Lahlou, Claire Escoffier, Pablo Pumares, and Paolo Ruspini. "L'immigration irrégulière subsaharienne à travers et vers le Maroc," *Cahiers de migrations internationales* 54F, Bureau International du Travail: Genève, 2002.

Barry, Tom. *Pushing Our Borders Out: Washington's Expansive Concept of Sovereignty and Security*. Silver City, NM: International Relations Center, 2005.

Bates, Diane C. "Environmental Refugees? Classifying Human Migrations Caused by Environmental Change." *Population and Environment* 23: 5 (May 2002): 465–477.

Battisti, David S., and Rosamond L. Naylor. "Historical Warnings of Future Food Insecurity Unprecedented Seasonal Heat." *Science* 323 (January 9, 2009): 240–244.

Bauer, Steffan. "Land and Water Scarcity as a Driver of Migration and Conflicts?" *Agriculture and Rural Development* 1 (2007): 7–9.

Belguendouz, Abdelkrim. *Le Maroc coupable d'emigration et de transit vers l'Europe*. Kénitra, Morocco: Boukili Impression, 2000.

———. *Le Maroc et la migration irrégulière: Une analyse sociopolitique*. Florence, Italy: Institut universitaire européen: Robert Schuman Centre for Advanced Studies, 2009.

Ben Jelloun, Tahar. *Leaving Tangier: A Novel*. Translated by Linda Coverdale. New York: Penguin, 2009.

Betz, Hans-Georg. "The New Politics of Resentment: Radical Right-Wing Political Parties in Western Europe." *Comparative Politics* 26: 4 (1993): 413–427.

Biasutti, Michela, and Alessandra Giannini. "Robust Sahel Drying in Response to Late 20th Century Forcings." *Geophysical Research Letters* 33: L11706 (2006).

Biermann, Frank. " 'Earth System Governance' as a Crosscutting Theme of Global Change Research." *Global Environmental Change* 17 (2007): 326–337.

Biermann, Frank, and Ingrid Boas. "Global Adaptation Governance: The Case of Protecting Climate Refugees." In *Global Climate Governance Beyond 2012: Architecture, Agency and Adaptation*, edited by Biermann et al. New York: Cambridge University Press, 2010, 255–269.

———. "Preparing for a Warmer World: Towards a Global Governance System to Protect Climate Refugees," *Global Environmental Politics* 10: 1 (February 2010): 60–88.

———. "Protecting Climate Refugees: The Case for a Global Protocol." *Environment* 50: 6 (November/December 2008): 8–16.

Biermann, Frank, and Klaus Dingwerth. "Global Environmental Change and the Nation State." *Global Environmental Politics* 4: 1 (2004): 1–23.

Biermann, Frank, Philipp Pattberg, and Fariborz Zelli, eds., "Global Climate Governance Beyond 2012: An Introduction." In *Global Climate Governance Beyond 2012: Architecture, Agency and Adaptation* (New York: Cambridge University Press, 2010), 1–12.

Black, Richard. *Environmental Refugees: Myth Or Reality?* Geneva, Switzerland: UNHCR Evaluation and Policy Analysis Unit, 2001.

———. "Fifty Years of Refugee Studies: From Theory to Policy." *International Migration Review* 35: 1 (Spring 2001): 57–78.

Bleibaum, Frauke. "Senegal." In *EACH-FOR: Environmental Change and Forced Migration Scenarios D.3.4. Synthesis Report*, edited by Andras Vag. Brussels, Belgium: European Commission, 2009, 44–45.

Bogardi, Janos, and Koko Warner. "Here Comes the Flood." *Nature Reports Climate Change* 138 (December 11, 2008).

Boko, Michel, Isabelle Niang, Anthony Nyong, Coleen Vogel, Andrew Githeko, Mahmoud Medany, Balgis Osman-Elasha, Ramadjita Tabo, and Pius Yanda. "Africa." In *Climate Change 2007: Impacts, Adaptation and Vulnerability: Contribution of Working Group II to the Fourth Assessment Report of the Intergovernmental Panel on Climate Change*, edited by M. L. Parry, O. F. Canziani, J. P. Palutikof, P. J. van der Linden, and C. E. Hanson. Cambridge, UK: Cambridge University Press, 2007, 433–467.

Borjas, G. J. "Economic Theory and International Migration." *International Migration Review* 23: 3 (1989).

———. "The New Economics of Immigration: Affluent Americans Gain, Poor Americans Lose." *Atlantic Monthly*, November 1996.

Brand, Laurie. *Citizens Abroad: Emigration and the State in the Middle East and North Africa*. New York: Cambridge University Press, 2006.

Bronen, Robin. "Forced Migration of Alaskan Indigenous Communities Due to Climate Change: Creating a Human Rights Response." Fairbanks, AK: University of Alaska Resilience and Adaptation Program, 2008.

Brubaker, Rogers. "Immigration, Citizenship, and the Nation-State in France and Germany: A Comparative Analysis." *International Sociology* 5: 4 (December 1990): 379–407.

Bull, Hedley. *The Anarchical Society: A Study of Order in World Politics*. New York: Columbia University Press, 1977.

Busby, Joshua W. "Who Cares about the Weather? Climate Change and U.S. National Security." *Security Studies* 17 (2008): 468–504.

Byravan, Sujatha, and Sudhir Chella Rajan. "The Ethical Implications of Sea-Level Rise Due to Climate Change." *Ethics and International Affairs* 24: 3 (2010): 239–260.

Campbell, Kurt, Jay Gulledge, J. R. McNeill, John Podesta, Peter Ogden, Leon Fuerth, R. James Woolsey, et al. *The Age of Consequences: The Foreign Policy and National Security Implications of Global Climate Change*. Washington, DC: Center for a New American Security and Center for Strategic and International Studies, 2007.

Campbell, Laura, W. Bradnee Chambers, Jerry Velasquez, and Shona E.H. Dodds. *Global Climate Governance: Inter-Linkages between the Kyoto Protocol and Other Multilateral Regimes*. Tokyo, Japan: United Nations University Institute of Advanced Studies and Global Environment Information Centre, 1999.

Cane, Mark. "The Evolution of El Niño, Past and Future." *Earth and Planetary Science Letters* 230 (2005): 227–240.

Cane, Mark, Stephen E. Zebiak, and Sean C. Dolan. "Experimental Forecasts of El Niño." *Nature* 321 (June 26, 1986).

Carling, Jørgen. "Migration Control and Migrant Fatalities at the Spanish-African Border." *International Migration Review* 41: 2 (2007): 316–343.

———. "Unauthorized Migration from Africa to Spain." *International Migration* 45: 4–37 (2007): 3–37.

Carmen, Commander Herbert E., USN, Christine Parthemore, and Will Rogers. *Broadening Horizons: Climate Change and the US Armed Forces*. Washington, DC: Center for a New American Security, 2010.

Cassarino, Jean-Pierre. "Theorising Return Migration: The Conceptual Approach to Return Migrants Revisited." *IJMS: International Journal on Multicultural Societies* 6: 2 (2004): 253–279.

Castles, Stephen. *Environmental Change and Forced Migration: Making Sense of the Debate*. Geneva, Switzerland: UNHCR Evaluation and Policy Analysis Unit, 2002.

———. "Towards a Sociology of Forced Migration and Social Transformation." *Sociology* 37: 13 (2003): 13–34.

Castles, Stephen, and Mark Miller. *The Age of Migration: International Population Movements in the Modern World*. 4th ed. New York: Palgrave Macmillan, 2010.

Chakrabarty, Dipesh. "The Climate of History." *Critical Inquiry* 35 (Winter 2009): 197–222.

Charef, Mohamed. "Geographical Situation as a Facilitator of Irregular Migration in Transit Countries—The Case of Tangier." Istanbul, Council of Europe, September 30–October 1, 2004.

Chimni, B. S., "The Geopolitics of Refugee Studies: A View from the South." *Journal of Refugee Studies* 11 (1998): 350–374.

Christensen, Jens Hesselbjerg, B. Hewitson, Anthony Chen, Xuejie Gao, Isaac Held, Richard Jones, Rupa Kumar Kolli, et al. "Regional Climate Projections." In *Climate Change 2007: The Physical Science Basis. Contribution of Working Group I to the Fourth Assessment Report of the Intergovernmental Panel on Climate Change*, edited by S. Solomon, D. Qin, M. Manning, Z. Chen, M. Marquis, K. B. Averyt, M. Tignor, and H. L. Miller. New York: Cambridge University Press, 2007, 847–879.

Christian Aid. *Human Tide: The Real Migration Crisis*. London: Christian Aid, 2007.

Cisneros, J. David. "Contaminated Communities: The Metaphor of 'Immigrant as Pollution' in Media Representations of Immigration." *Rhetoric and Public Affairs* 11: 4 (2008): 569–602.

Clapp, Jennifer, and Peter Dauvergne. *Paths to a Green World: The Political Economy of the Global Environment*. Cambridge, MA: MIT Press, 2005.

Collyer, Michael. *States of Insecurity: Consequences of Saharan Transit Migration*. Oxford, England: University of Oxford Centre on Migration, Policy and Society, 2006.

Conca, Ken. "Rethinking the Ecology-Sovereignty Debate." *Millennium* 23: 3 (1994): 701–711.

Cooper, Frederick. *Colonialism in Question: Theory, Knowledge, History*. Berkeley: University of California Press, 2005.

Cordell, Dennis D., Joel W. Gregory, and Victor Piché. *Hoe and Wage: A Social History of a Circular Migration System in West Africa*. Boulder, CO: Westview, 1996.

Cornelius, Wayne. "Death at the Border: Efficacy and Unintended Consequences of U.S. Immigration Control Policy." *Population and Development Review* 27: 4 (2001): 661–685.

———. "Spain: The Uneasy Transition from Labor Exporter to Labor Importer." In *Controlling Immigration: A Global Perspective*, edited by Wayne Cornelius, Philip Martin, and James Hollifield. 2nd ed. Stanford, CA: Stanford University Press, 2004.

Coultin, Susan Bibler. "Being En Route." *American Anthropologist* 107: 2 (2005): 195–206.

Council of the European Union, "Joint Statement European Union-Morocco Summit, Granada, 7 March 2010." 7220/10 (Presse 54). Brussels: Europa, available at www.consilium.europa.edu/newsroom.

Crichton, Michael. *State of Fear*. New York: Avon Books, 2004.

Dabelko, Geoffrey. "Avoid Hyperbole, Oversimplification When Climate and Security Meet." *Bulletin of the Atomic Scientists*, August 24, 2009.

———. "Planning for Climate Change: The Security Community's Precautionary Principle." *Climate Change* 96 (2009): 13–21.

Dalby, Simon. *Security and Environmental Change*. Malden, MA: Polity, 2009.

Davidson, Ogunlade, Kirsten Halsnæs, Saleemul Huq, Marcel Kok, Bert Metz, Youba Sokona, and Jan Verhagen. "The Development and Climate Nexus: The Case of Sub-Saharan Africa." *Climate Policy* 3S1 (2003): S97–S113.

Davis, Diana K. *Resurrecting the Granary of Rome: Environmental History and French Colonial Expansion in North Africa*. Athens: Ohio University Press, 2007.

de Haas, Hein. *Irregular Migration from West Africa to the Maghreb and the European Union: An Overview of Recent Trends*. Geneva: International Organization for Migration, 2005.

———. "Morocco's Migration Experience: A Transitional Perspective." *International Migration* 45: 4 (2007): 39–70.

———. *Morocco's Migration Transition: Trends, Determinants and Future Scenarios*. Nijmegen, The Netherlands: Centre for International Development Issues, Radboud University, 2005.

———. *The Myth of Invasion: Irregular Migration from West Africa to the Maghreb and the European Union* (Oxford, England: International Migration Institute, 2007).

de Sherbinin, Alex, Andrew Schiller, and Alex Pulsipher. "The Vulnerability of Global Cities to Climate Hazards." *Environment and Urbanization* 19 (2007): 39–64.

de Sherbinin, Alex, Koko Warner, and Charles Ehrhart. "Casualties of Climate Change." *Scientific American* 304: 1, 2011, 64–71.

De Souza, Roger-Mark. *Focus on Population, Environment, and Security*. Washington, DC: Environmental Change and Security Program of the Woodrow Wilson Center, 2009.

de Wit, Maarten, and Jacek Stankiewicz. "Changes in Water Supply Across Africa with Predicted Climate Change." *Science* 311 (2006): 1917–1921.

Denoeux, Guilain, and Abdeslam Maghraoui. "King Hassan's Strategy of Political Dualism." *Middle East Policy* 5: 4 (January 1998): 104–130.

Detraz, Nicole, and Michele Betsill. "Climate Change and Environmental Security: For Whom the Discourse Shifts." *International Studies Perspectives* 10 (2009): 303–320.

Deudney, Daniel. "The Case against Linking Environmental Degradation and National Security." *Millennium* 19: 3 (1900): 461–476.

Deutsch, Karl. *Political Community and the North Atlantic Area: International Organization in the Light of Historical Experience*. Princeton, NJ: Princeton University Press, 1957.

Dietz, Mary. "Trapping the Prince: Machiavelli and the Politics of Deception." *American Political Science Review* 80 (1986): 777–799.

Düvall, Franck. *Crossing the Fringes of Europe: Transit Migration in the EU's Neighbourhood*. Centre on Migration, Policy and Society: University of Oxford Working Paper No. 33, 2006.

Dyer, Gwynne. *Climate Wars*. London: Oneworld, 2010.

Eakin, Hallie, and Amy Lynd Luers. "Assessing the Vulnerability of Social-Environmental Systems." *Annual Review of Environmental Resources* 31 (2006): 365–394.

Earth Institute and United Nations Development Program. *The Millennium Villages Project: Annual Report 2008*. New York: Columbia University, 2008.

Ehrenreich, Barbara, and Arlie Hochschild, eds. *Global Woman: Nannies, Maids and Sex Workers in the New Economy*. New York: Holt, 2004.

Elbe, Stefan. "Should HIV/AIDS Be Securitized? The Ethical Dilemmas of Linking HIV/AIDS and Security." *International Studies Quarterly* 50: 1 (March 2006): 119–144.

El-Hinnawi, Essam. *Environmental Refugees*. Nairobi, Kenya: United Nations Environment Program, 1985.

Elmadmad, Khadija. *Migration irrégulière et migration illégale: L'exemple des migrants subsahariens au Maroc*. Florence, Italy: Robert Schuman Centre for Advanced Studies— European University Institute, 2008.

Ennaji, Moha, and Fatimi Sadiqi. *Migration and Gender in Morocco: The Impact of Migration on the Women Left Behind*. Trenton, NJ: Red Sea Press, 2008.

Ennaji, Mohammed. *Soldats, domestiques et concubines: L'esclavage au Maroc au XIXe siècle*. Casablanca: Editions EDDIF, 1997.

Euro-African Ministerial Conference on Migration and Development. "Rabat Plan of Action of the Euro-African Ministerial Conference on Migration and Development." Rabat, Morocco, 2006.

European Commission. *Aeneas Programme: Overview of Projects Funded 2004–2006*. Brussels, Belgium: Europe Aid Programme of the European Commission, 2008.

———. "European Commission and Libya Agree to a Migration Cooperation Agenda during High Level Visit to Boost EU-Libya Relations." Brussels, Belgium, October 5, 2010, europa.eu/rapid/pressReleasesAction.do?reference=MEMO/10/472.

Evans, Peter B., Harold K. Jacobson, and Robert Putnam, eds. *Double-Edged Diplomacy: International Bargaining and Domestic Politics*. Berkeley: University of California Press, 1993.

Feliu Martínez, Laura. "Les migrations en transit au Maroc: Attitudes et comportement de la société civile face au phénomène," *L'Année du Maghreb* 5 (2009): 343–362.

Findley, Sally, Sadio Traoré, Dieudonné Ouedraogo, and Sekouba Diarra. "Emigration from the Sahel." *International Migration* 33: 3–4 (1995): 459–520.

Finnemore, Martha. "Constructing Norms of Humanitarian Intervention." In *The Culture of National Security: Norms and Identity in World Politics*, edited by Peter Katzenstein. New York: Columbia University Press, 1996, 153–185.

Foley, Jonathan. "Boundaries for a Healthy Planet." *Scientific American* 302: 4, 2010, 54–57.

Freeman, Gary. "Immigrant Labour and Working-Class Politics: The French and British Experience." *Comparative Politics* 11: 1 (1978): 25–41.

Friedman, Lisa. How Will Climate Refugees Impact National Security? *Scientific American*, March 23, 2009.

Friedman, Thomas. *The Lexus and the Olive Tree*. New York: Farrar, Straus & Giroux, 1999.

Funk, Chris, Michael D. Dettinger, Joel C. Michaelsen, James P. Verdin, Molly E. Brown, Mathew Barlow, and Andrew Hoell. "Warming of the Indian Ocean Threatens Eastern and Southern African Food Security but Could Be Mitigated by Agricultural Development." *Proceedings of the National Academy of Sciences* 105: 32 (August 12, 2008): 11081–11086.

Gardiner, Stephen. "A Perfect Moral Storm: Climate Change, Intergenerational Ethics, and the Problem of Corruption." In *Political Theory and Global Climate Change*, edited by Steve Vanderheiden. Cambridge, MA: MIT Press, 2008, 25–42.

German Advisory Council on Global Change. *World in Transition: Climate Change as a Security Risk*. Berlin, Germany, 2007.

Giannini, Alessandra, Michela Biasutti, Isaac M. Held, and Adam H. Sobel. "A Global Perspective on African Climate." *Climatic Change* 90 (2008): 359–383.

Giannini, Alessandra, R. Saravanan, and P. Chang. "Oceanic Forcing of Sahel Rainfall on Interannual to Interdecadal Time Scales." *Science* 302 (November 7, 2003): 1027–1030.

Gillespie, Richard. *Spain and Morocco: Towards a Reform Agenda?* Madrid: Fundación para las relaciones internacionales y el diálogo exterior, 2005.

Goldenberg, Stanley, Christopher Landsea, Alberto M. Mestas-Nuñez, and William M. Gray. "The Recent Increase in Atlantic Hurricane Activity: Causes and Implications." *Science* 293 (2001): 474–479.

Goldschmidt, Elie. "Storming the Fences: Morocco and Europe's Anti-Migration Policy." *Middle East Report Online* 239 (2006), available at www.merip.org/mer/mer239/goldschmidt.html.

Gordimer, Nadine. "Once upon a Time." In *Jump and Other Short Stories*. New York: Penguin, 1992.

Gore, Al. *Earth in the Balance*. New York: Houghton Mifflin, 1992.

———. "Introduction." In *Silent Spring*, by Rachel Carson. New York: Houghton Mifflin, 1994.

Gränzer, Sieglinde. "Changing Human Rights Discourse: Transnational Advocacy Networks in Tunisia and Morocco." In *The Power of Human Rights: International Norms and Domestic Change*, edited by Thomas Risse, Stephen C. Ropp and Kathryn Sikkink. New York: Cambridge University Press, 1999, 109–133.

Gray, Clark L. "Environment, Land, and Rural Out-Migration in the Southern Ecuadorian Andes." *World Development* 37: 2 (2009): 457–468.

———. *Environmental Refugees or Economic Migrants?* Washington, DC: Population Reference Bureau, 2010, available at www.prb.org.

Guild, Elspeth. *Security and Migration in the 21st Century*. Malden, MA: Polity Press, 2009.

Guiraudon, Virginie, and Gallya Lahav. "A Reappraisal of the State Sovereignty Debate: The Case of Migration Control." *Comparative Political Studies* 33: 2 (2000): 163–195.

Gunn, Lee. "Introduction." In *Climate Security Initiative: Climate Security Index*. Washington, DC: American Security Project, 2009.

Gutelius, David. "Islam in Northern Mali and the War on Terror." *Journal of Contemporary African Studies* 25: 1 (2007): 59–76.

Haas, Peter. "Constructing Environmental Conflicts from Resource Scarcity." *Global Environmental Politics* 2: 1 (February 2002): 1–11.

———. *Saving the Mediterranean: The Politics of International Environmental Cooperation*. New York: Columbia University Press, 1990.

———. "Social Constructivism and the Evolution of Multilateral Environmental Governance." In *Globalization and Governance*, edited by Aseem Prakash and Jeffrey A. Hart. New York: Routledge, 1999, 103–133.

Haas, Peter, Robert Keohane, and Marc Levy, eds. *Institutions for the Earth: Sources of Effective International Environmental Protection*. Cambridge, MA: MIT Press, 1993.

Haddad, Emma. *The Refugee in International Society: Between Sovereigns*. New York: Cambridge University Press, 2008.

Hammoudi, Abdellah. *Master and Disciple: The Cultural Foundations of Moroccan Authoritarianism*. Chicago: Chicago University Press, 1997.

Hansen, James. "Defusing the Global Warming Time Bomb." *Scientific American*, March 2004, 68–77.

Hansen, James, Larissa Nazarenko, Reto Ruedy, Makiko Sato, Josh Willis, Anthony Del Genio, Dorothy Koch, et al. "Earth's Energy Imbalance: Confirmation and Implications." *Science* 308 (June 3, 2005): 1431–1435.

Hartmann, Betsy. "Rethinking Climate Refugees and Climate Conflict: Rhetoric, Reality and the Politics of Policy Discourse." *Journal of International Development* 22 (2010): 233–246.

Hein, Lars, and Nico De Ridder. "Desertification in the Sahel: A Reinterpretation." *Global Change Biology* 12 (2006): 751–758.

Heisler, Martin O., and Zig Layton-Henry. "Migration and the Links between Social and Societal Security." In *Identity, Migration and the New Security Agenda in Europe*, edited by Ole Wæver, Barry Buzan, Morten Kelstrup and Pierre Lemaitre. New York: St. Martin's, 1993, 148–166.

Henry, Sabine, Paul Boyle, and Eric F. Lambin. "Modelling Inter-Provincial Migration in Burkina Faso, West Africa: The Role of Socio-Demographic and Environmental Factors." *Applied Geography* 23 (2003): 115–136.

Henry, Sabine, Bruno Schoumaker, and Cris Beauchemin. "The Impact of Rainfall on the First Out-Migration: A Multi-Level Event-History Analysis in Burkina Faso." *Population and Environment* 25: 5 (May 2004): 423–460.

Herrero, Sergio Tirado. "Desertification and Environmental Security: The Case of Conflicts between Farmers and Herders in the Arid Environments of the Sahel." In *Desertification in the Mediterranean Region: A Security Issue*, edited by William G. Kepner, José L. Rubio, David A. Mouat, and Fausto Pedrazzini. Dordrecht, The Netherlands: Springer, 2006, 109–132.

Hollifield, James F. "The Politics of International Migration: How Can We 'Bring the State Back In?' " In *Migration Theory: Talking Across Disciplines*, edited by C. B. Brettell and J. F. Hollifield. 2nd ed. New York: Routledge, 2007, 183–237.

Homer-Dixon, Thomas. "Environmental Scarcities and Violent Conflict: Evidence from Cases." *International Security* 19: 1 (1994): 5–40.

———. "On the Threshold: Environmental Changes as Causes of Acute Conflict." *International Security* 16: 2 (1991): 76–116.

Hondagneu-Sotelo, Pierette. *Gendered Transitions: Mexican Experiences of Immigration.* Berkeley: University of California Press, 1995.

Hood, David. *Fatal Climate*. London: Phoenix, 2001.

Hufbauer, Clyde, and Gustavo Vega-Cánovas. "Whither NAFTA: A Common Frontier?" In *The Rebordering of North America: Integration and Exclusion in a New Security Context*, edited by Peter Andreas and Thomas Biersteker. New York: Routledge, 2003, 128–152.

Hunter, Lori M., and Emmanuel David. *Climate Change and Migration: Considering the Gender Dimensions.* POP2009-13. University of Colorado at Boulder: Institute of Behavioral Science, 2009.

Huntington, Samuel. "The Clash of Civilizations?" *Foreign Affairs* 72: 3 (1993): 22–49.

———. *Who Are We? The Challenges to America's National Identity.* New York: Simon & Schuster, 2005.

International Centre for Migration Policy Development, Europol, and Frontex. *Arab and European Partner States Working Document on the Join Management of Mixed Migration Flows.* Luxembourg: Office for Official Publications of the European Communities, 2007.

International Centre for Migration Policy Development and Ministère de l'immigration, de l'identité et du developpement solidaire of the République Francaise. *Dialogue on Mediterranean Transit Migration and i-Map Expert Meeting, Paris, 15–16 December.* Vienna, Austria: ICMPD, 2008.

International Centre for Migration Policy Development and Syrian Arab Republic Ministry of Interior. *Dialogue on Mediterranean Transit Migration and i-MAP Expert Meeting, Damascus, Syria, 30 June–1 July.* Vienna, Austria: ICMPD, 2009.

International Organization for Migration. *Climate Change, Environmental Degradation and Migration: Addressing Vulnerabilities and Harnessing Opportunities: Discussion*

Note on Migration and the Environment MC/INF/288. Geneva, Switzerland: International Organization for Migration, 2008.

International Research Institute for Climate and Society, Global Climate Observing System, United Kingdom's Department for International Development, and UN Economic Commission for Africa. *A Gap Analysis for the Implementation of the Global Climate Observing System Programme in Africa*. IRI-TR/06/1. Palisades, NY: IRI, 2006.

Iskander, Natasha. *Creative State: Forty Years of Migration and Development Policy in Morocco and Mexico*. Ithaca, NY: Cornell University Press, 2010.

Ivanova, Maria. "Moving Forward by Looking Back: Learning from UNEP's History." In *Green Planet Blues*, ed. Ken Conca and Geoffrey Dabelko. 4th ed. Boulder, CO: Westview, 2010, 143–160.

Jackson, Robert, and Carl Rosberg. "Why Africa's Weak States Persist: The Empirical and the Juridical in Statehood." *World Politics* 35: 1 (1982): 3–32.

Jacobsen, Karen. "Refugees and Asylum Seekers in Urban Areas: A Livelihoods Perspective." *Journal of Refugee Studies* 19: 3 (2006): 373–286.

Jones, Reece. "Geopolitical Boundary Narratives, the Global War on Terror and Border Fencing in India." *Transactions of the Institute of British Geographers* 34 (2009): 290–304.

Joppke, Christian. *Immigration and the Nation-State: The United States, Germany, and Great Britain*. New York: Oxford University Press, 1999.

Jourde, Cédric. "Constructing Representations of the 'Global War on Terror' in the Islamic Republic of Mauritania." *Journal of Contemporary African Studies* 25: 1 (January 2007): 77–100.

Junger, Sebastian. *The Perfect Storm*. New York: Norton, 1997.

Kaplan, Robert. "The Coming Anarchy." *Atlantic Monthly*, February 1994, 44–77.

Karns, Margaret P., and Karen A. Mingst. *International Organizations: The Politics and Processes of Global Governance*. 2nd ed. Boulder, CO: Lynne Rienner, 2009.

Keck, Margaret, and Kathryn Sikkink. *Activists beyond Borders: Advocacy Networks in International Politics*. Ithaca, NY: Cornell University Press, 1998.

Keohane, Robert O., and Joseph S. Nye. *Power and Interdependence*. 2nd ed. Boston, MA: Little Brown, 1989.

Keohane, Robert O., and David G. Victor. *The Regime Complex for Climate Change*. Cambridge, MA: Harvard Kennedy School Project on International Climate Agreements, 2010.

Kernerman, Gerald. "Refugee Interdiction before Heaven's Gate." *Government and Opposition* 43: 2 (2008): 230–248.

Khader, Bichara. "Immigration and the Euro-Mediterranean Partnership." In *The Euro-Mediterranean Partnership: Assessing the First Decade*, edited by Haizam Amirah Fernández and Richard Youngs. Barcelona: Fundación para las realaciones internacionales y el dialogo exterior, 2005, 83–92.

Kibreab, Gaim. "Environmental Causes and Consequences of Migration: A Search for the Meaning of 'Environmental Refugee.'" *Disasters* 21: 1 (1997): 20–38.

Kimball, Ann. *The Transit State: A Comparative Analysis of Mexican and Moroccan Immigration Policies*. Center for Iberian and Latin American Studies and Center for Comparative Immigration Studies, University of California-San Diego, 2007.

Kindleberger, Charles. *The World in Depression, 1929–39*. Berkeley: University of California Press, 1973.

King, Leslie. "Ideology, Strategy and Conflict in a Social Movement Organization: The Sierra Club Immigration Wars." *Mobilization: An International Quarterly* 13: 1 (2008): 45–61.

Kingdom of Spain and Kingdom of Morocco. *Treaty of Friendship, Good-Neighbourliness and Cooperation —Signed in Rabat, Morocco*, no. 1717, I–29862 (July 4, 1991), available at www.untreaty.un.org/unts/120001_144071/3/1/00001779.pdf.

Klare, Michael. *Rising Powers, Shrinking Planet: The New Geopolitics of Energy*. New York: Metropolitan, 2008.

Klare, Michael, and Daniel Volman. "The African 'Oil Rush' and US National Security." *Third World Quarterly* 27: 4 (2006): 609–628.

Kniveton, Dominic, Kerstin Schmidt-Verkerk, Christopher Smith, and Richard Black. *Climate Change and Migration: Improving Methodologies to Estimate Flows*. Geneva, Switzerland: International Organization for Migration, 2008.

Kolmannskog, Vikram Odedra. *Future Flood Of Refugees: A Comment on Climate Change, Conflict and Forced Migration*. Oslo, Norway: Norwegian Refugee Council, 2008.

Koslowski, Rey. "The Mobility Money Can Buy: Human Smuggling and Border Control in the European Union." In *The Wall around the West: State Borders and Immigration Controls in North America and Europe*, edited by Peter Andreas and Timothy Snyder. Lanham, MD: Rowman & Littlefield, 2000, 203–218.

Krasner, Stephen. *Sovereignty: Organized Hypocrisy*. Princeton, NJ: Princeton University Press, 1999.

Kump, L. R., J. F. Kasting, and R. Crane. *The Earth System*. Upper Saddle River, NJ: Prentice-Hall, 2004.

Kyle, David, and Rey Koslowski, eds. *Global Human Smuggling: Comparative Perspectives*. Baltimore, MD: Johns Hopkins University Press, 2001.

Laczko, Frank, and Christine Aghazarm, eds. *Migration, Environment and Climate Change: Assessing the Evidence*. Geneva: International Organization for Migration, 2010.

Lake, David A. "British and American Hegemony Compared: Lessons for the Current Era of Decline." In *International Political Economy: Perspectives on Global Power and Wealth*, edited by Jeffry Frieden and David A. Lake. 3rd. ed. New York: St. Martin's, 2000, 148–166.

Lalami, Laila. *Hope and Other Dangerous Pursuits*. Chapel Hill, NC: Algonquin Books, 2005.

Landau, Loren, and Aurelia Kazadi Wa Kabwe-Segatti, *Human Development Impacts of Migration: South Africa Case Study*. United Nations Development Programme Research Paper 2009/5.

Larramendi, Miguel Hernando de. *Las relaciones con Marruecos tras los atentados del 11 de Marzo*. Madrid: Real Instituto Elcano de Estudios Internacionales y Estratégicas, 2004.

Layachi, Azzedine. "State-Society Relations and Change in Morocco." In *Economic Crisis and Political Change in North Africa*, edited by Azzedine Layachi. Westport, CT: Praeger, 1998, 89–106.

Leach, Melissa, and James Fairhead. "Challenging Neo-Malthusian Deforestation Analyses in West Africa's Dynamic Forest Landscapes." *Population and Development Review* 26: 1 (March 2000): 17–43.

Lenton, Timothy M., Hermann Held, Elmar Kriegler, Jim W. Hall, Wolfgang Lucht, Stefan Rahmstorf, and Hans Joachim Schellnhuber. "Tipping Elements in the Earth's Climate System." *Proceedings of the National Academy of Sciences* 105: 6 (February 12, 2008): 1786–1793.

Levy, Marc. "Is the Environment a National Security Issue?" *International Security* 20: 2 (Fall 1995): 35–62.

Linklater, Andrew, and Hidemi Suganami. *The English School of International Relations: A Contemporary Reassessment.* London: Cambridge University Press, 2005.

Loescher, Gil. *Beyond Charity: International Cooperation and the Global Refugee Crisis.* New York: Oxford University Press, 1993.

López-García, Bernabé. "Foreign Immigration Comes to Spain: The Case of the Moroccans." In *New European Identity and Citizenship,* edited by Rémy Leveau, Khadija Mohsen-Finan, and Catherine Withol de Wenden. Aldershot, UK: Ashgate, 2002, 49–68.

Luseno, Winnie, John McPeack, Christopher Barrett, Peter Little, and Getachew Gebru. "Assessing the Value of Climate Forecast Information for Pastoralists: Evidence from Southern Ethiopia and Northern Kenya." *World Development* 31: 9 (2003): 1477–1494.

Lutterbeck, Derek. "Migrants, Weapons and Oil: Europe and Libya After the Sanctions." *Journal of North African Studies* 14: 2 (2009): 169–184.

———. "Policing Migration in the Mediterranean." *Mediterranean Politics* 11: 1 (March 2006): 59–82.

Maghraoui, Abdeslam. "From Symbolic Legitimacy to Democratic Legitimacy: Monarchic Rule and Political Reform in Morocco." *Journal of Democracy* 12: 1 (2001): 73–86.

Marfleet, Philip. *Refugees in a Global Era.* New York: Palgrave Macmillan, 2006.

Martin, Susan F. "Climate Change and International Migration." Background paper on Climate Change and Migration for the German Marshall Fund of the United States, available at www.gmfus.org, June 2010.

———. "Climate Change, Migration, and Governance." *Global Governance* 16 (2010): 397–414.

———. "Managing Environmentally Induced Migration." In *Migration, Environment and Climate Change: Assessing the Evidence,* edited by Frank Laczko and Christine Aghazarm. Geneva: International Organization for Migration, 2010, 353–384.

———. *Refugee Women.* 2nd ed. Lanham, MD: Lexington Books, 2004.

Martin, Susan Forbes et al. *The Uprooted: Improving Humanitarian Responses to Forced Migration.* Lanham, MD: Rowman & Littlefield, 2005.

Massey, Douglas, Joaquin Arango, Hugo, Graeme, Kouaaouci, Ali, Pellegrino, Adela, and J. Edward Taylor. "Theories of International Migration: A Review and Appraisal." *Population and Development Review* 19: 3 (1993): 431–466.

Massey, Douglas, William Axinn, and Dirgha Ghimire. *Environmental Change and Out-Migration: Evidence from Nepal.* University of Michigan Institute for Social Research: Population Studies Center Report 07-615, 2007.

McMurray, David. *In and Out of Morocco: Smuggling and Migration in a Frontier Boomtown.* Minneapolis: University of Minnesota, 2001.

Meadows, Donella H., Dennis L. Meadows, Jørgen Randers, and William W. Behrens III. "The Limits to Growth." In *Green Planet Blues,* edited by Ken Conca and Geoffrey D. Dabelko. 4th ed. Boulder, CO: Westview, 2010, 25–29.

Mearsheimer, John. *The Tragedy of Great Power Politics.* New York: Norton, 2001.

Mohammed VI. "Speech to the 2009 Meeting of the International Union for the Scientific Study of Population." Marrakech, Morocco, September 27, 2009, available at www.map.ma.

———. "Speech to the 2nd EU-Africa Summit Hosted by Lisbon." Lisbon, Portugal, December 8, 2009, available at www.map.ma.

———. "Speech to the 3rd EU-Africa Summit hosted by Libya." Tripoli, Libya, November 29, 2010, available at www.map.ma.

Monimart, Marie. *Femmes du Sahel*. Paris: Karthala, 1989.

Mooney, Chris. *Unscientific America: How Science Illiteracy Threatens our Future*. New York: Basic Books, 2009.

Moorehead, Caroline. *Human Cargo: A Journey among Refugees*. New York: Henry Holt, 2005.

Morganthau, Hans. *Politics among Nations: The Struggle for Power and Peace*. New York: Knopf, 1960.

Mortimore, Michael. "Adapting to Drought in the Sahel: Lessons for Climate Change." *WIRES Climate Change* 1 (2010): 134–143.

Myers, Norman. "Environmental Refugees." *Population and Environment* 19 (1997): 167–182.

———. "Environmental Refugees: An Emergent Security Issue." Paper presented at the 13th Economic Forum, Prague, Czechoslovakia, 2005.

———. "Environmental Refugees: A Growing Phenomenon of the 21st Century." *Philosophical Transactions of the Royal Society* 356 (2001).

National Intelligence Council. *Global Trends 2025: A Transformed World*. Washington, DC: US Government Printing Office, 2008.

National Research Council. *Abrupt Climate Change: Inevitable Surprises*. Washington, DC: National Academy Press, 2002.

Newman, David. "Boundaries." In *A Companion to Political Geography*, edited by John Agnew, Katharyne Mitchell and Gerard Toal. Malden, MA: Blackwell, 2003, 123–137.

Nordås, Ragnhild and Nils Petter Gleditsch. "Climate Change and Conflict." *Political Geography* 26: 6 (2007): 627–638.

Olsson, L., L. Eklundh, and J. Ardö. "A Recent Greening of the Sahel: Trends, Patterns and Potential Causes." *Journal of Arid Environments* 63 (2005): 556–566.

O'Neill, Brian, Simone Pulver, Stacy VanDeveer, and Yaakov Garb. "Where Next with Global Environmental Scenarios?" *Environmental Research Letters* 3: 4 (2008): 1–4.

Oreskes, Naomi. "The Scientific Consensus on Climate Change: How Do We Know We're Not Wrong?" In *Climate Change: What It Means for Us, Our Children, and Our Grandchildren*, edited by Joseph F. C. Dimento and Pamela Doughman. New York: Cambridge University Press, 2007.

Oreskes, Naomi, and Erik Conway. *Merchants of Doubt: How a Handful of Scientists Obscured the Truth on Issues from Tobacco Smoke to Global Warming*. London: Bloomsbury Press, 2010.

Orlove, Benjamin S., John C. H. Chiang, and Mark A. Cane. "Forecasting Andean Rainfall and Crop Yield from the Influence of El Niño on Pleiades Visibility." *Nature* 403: 6 (2000): 68–71.

Orwell, George. "Politics and the English Language." In *A Collection of Essays*. Orlando, FL: Harcourt, 1970.

Panda, Architesh. "Climate Refugees: Implications for India." *Economic & Political Weekly* 45: 20 (May 15, 2010): 76–79.

Paskal, Cleo. *Global Warring: How Environmental, Economic and Political Crises Will Redraw the World Map*. London: Palgrave Macmillan, 2010.

———. "How Climate Change Is Pushing the Boundaries of Security and Foreign Policy." 07/01, Energy. Environment and Development Programme, Royal Institute of International Affairs, Chatham House, London, June 2007.

Paul, Bimal Kanti. "Evidence against Disaster-Induced Migration: The 2004 Tornado in North-Central Bangladesh." *Disasters* 29: 4 (2005): 370–385.

Perch-Nielson, Sabine, Michèle B. Bättig, and Dieter Imboden. "Exploring the Link between Climate Change and Migration." *Climatic Change* 91 (2008): 375–393.

Permanent Select Committee on Intelligence and House Select Committee on Energy Independence and Global Warming. "Testimony by Thomas Fingar, Deputy Director for National Intelligence, on the National Security Implications of Global Climate Change to 2030." Washington, DC: U.S. House of Representatives, June 25, 2008.

Pickerill, Emily. "Informal and Entrepreneurial Strategies among Sub-Saharan Migrants in Morocco." *Journal of North African Studies* (forthcoming).

Pielke, Roger. *Honest Broker: Making Sense of Science in Policy and Politics.* New York: Cambridge University Press, 2009.

Piguet, Etienne. *Climate Change and Forced Migration.* Geneva, Switzerland: UN High Commission for Refugees Policy Development and Evaluation Service, 2008.

Piore, Michael. *Birds of Passage: Migrant Labor in Industrial Societies.* Princeton, NJ: Princeton University Press, 1979.

Powers, Holiday. "The Challenges of Maintaining Local Identity in International Biennale Exhibitions: Lessons from the 3rd AiM Arts in Marrakech Biennale." Paper presented at the 2010 AIMS Conference, June 27, 2010, University of Oran, Algeria.

Prince, Stephen D., Konrad J. Wessels, Compton J. Tucker, and Sharon E. Nicholson. "Desertification in the Sahel: A Reinterpretation of a Reinterpretation." *Global Change Biology* 13 (2007): 1308–1313.

Pulver, Simone, and Stacy VanDeveer. " 'Thinking about Tomorrows': Scenarios, Global Environmental Politics, and Social Science Scholarship." *Global Environmental Politics* 9: 2 (May 2009): 1–13.

Rahmstorf, Stefan. "Ocean Circulation and Climate during the Past 120,000 Years." *Nature* 419 (September 12, 2002): 207–214.

Rain, David. *Eaters of the Dry Season: Circular Labor Migration in the West African Sahel.* Boulder, CO: Westview, 1999.

Raleigh, Clionadh. "Political Marginalization, Climate Change and Conflict in African Sahel States." *International Studies Review* 12 (2010): 69–86.

Ratha, Dilip, and Xu, Zhimei. "World Bank Migration and Remittances Factbook." Available at www.worldbank.org/prospects/migrationandremittances2009.

Rawls, John. *The Law of Peoples.* Cambridge, MA: Harvard University Press, 1999.

———. *A Theory of Justice.* Cambridge, MA: Harvard University Press, 1971.

Raynaut, Claude. "Societies and Nature in the Sahel: Ecological Diversity and Social Dynamics." *Global Environmental Change* 11 (2001): 9–18.

Reij, C., G. Tappan, and A. Belemvire. "Changing Land Management Practices and Vegetation on the Central Plateau of Burkina Faso (1968–2002)." *Journal of Arid Environments* 63 (2005): 642–569.

Reynolds, James F., D. Mark Stafford Smith, Eric F. Lambin, B.L. Turner, Michael Mortimore, Simon P.J. Batterbury, Thomas E. Downing, et al. "Global Desertification: Building a Science for Dryland Development." *Science* 316 (May 11, 2007): 847–851.

Riaz, Ali. "Bangladesh." In *Climate Change and National Security: A Country-Level Analysis*, edited by Daniel Moran. Washington, DC: Georgetown University Press, 2010, 103–114.

Ribas-Mateos, Natalia. "Female Birds of Passage: Leaving and Settling in Spain." In *Gender and Migration in Southern Europe: Women on the Move*, edited by Floya Anthias and Gabriella Lazaridis. New York: Berg, 2000, 173–197.

Rockström, Johan, Will Steffan, Kevin Noone, Asa Persson, F. Stuart Chapin III, Eric F. Lambin, Timothy M. Lenton, et al. "A Safe Operating Space for Humanity." *Nature* 461 (September 24, 2009): 472–475.

Rosenzweig, Cynthia, Gino Casassa, David Karoly, Anton Imeson, Chunzhen Liu, Annette Menzel, Samuel Rawlins, Terry Root, Bernard Seguin, and Piotr Tryjanowski. "Assessment of Observed Changes and Responses in Natural and Managed Systems." In *Climate Change 2007: Impacts, Adaptation and Vulnerability: Contribution of Working Group II to the Fourth Assessment Report of the Intergovernmental Panel on Climate Change*, edited by M. L. Parry, O. F. Canziani, J. P. Palutikof, P. J. van der Linden, and C. E. Hanson. Cambridge, UK: Cambridge University Press, 2007, 79–131.

Rousseau, Jean-Jacques. *On the Social Contract.* Translated by Judith R. Masters, edited by Roger D. Masters. New York: St. Martin's, 1978.

Rudolph, Christopher. "Security and the Political Economy of International Migration." *American Political Science Review* 97: 4 (2003): 603–620.

Sachs, Jeffrey. "Climate Change Refugees: As Global Warming Tightens the Availability of Water, Prepare for a Torrent of Forced Migrations." *Scientific American*, June 1, 2007.

———. *Common Wealth: Economics for a Crowded Planet.* New York: Penguin, 2008.

Sadiqi, Fatima. *Migration-Related Institutions and Policies in Morocco.* European University Institute, Florence, Italy: Euro-Mediterranean Consortium for Applied Research on International Migration (CARIM), 2004.

Sahin, Zeynap. "Policy Changes in the Immigration Controls of States after 1990s: The Case of Turkey." International Studies Association Annual Meeting, New York, March 23, 2009.

Salehyan, Idean. "The New Myth about Climate Change." *Foreign Policy*, August 2007, available at www.foreignpolicy.com.

Salime, Zakia. "The War on Terrorism: Appropriation and Subversion by Moroccan Women." *Signs: Journal of Women in Culture and Society* 33: 1 (2007): 1–24.

Salter, Mark B. "Passports, Mobility, and Security: How Smart Can the Border Be?" *International Studies Perspectives* 5 (2004): 71–91.

Sassen, Saskia. *Globalization and Its Discontents: Essays on the New Mobility of People and Money.* New York: Free Press, 1998.

———. *Losing Control? Sovereignty in an Age of Globalization.* New York: Columbia University Press, 1996.

———. "Whose City Is It? Globalization and the Formation of New Claims." In *Globalization and Its Discontents: Essays on the New Mobility of People and Money.* New York: Free Press, 1998.

Schipper, E. Lisa F. "Conceptual History of Adaptation in the UNFCCC Process." *Review of European Community and International Environmental Law* 15: 1 (2006): 82–92.

Schmidt, Gavin A. "The Physics of Climate Modeling." *Physics Today* (January 2007): 72–73.

Schneider, Stephen. *Science as a Contact Sport: Inside the Battle to Save the Earth's Climate.* Washington, DC: National Geographic, 2009.

Schwartz, Peter, and Doug Randall. *An Abrupt Climate Change Scenario and Its Implications for United States National Security.* New York: Global Business Network, 2003.

Seager, Jody. *Earth Follies: Coming to Feminist Terms with the Global Environmental Crisis.* New York: Routledge, 1993.

Seager, Richard, Mingfang Ting, Isaac Held, Yochanan Kushnir, Jian Lu, Gabriel Vecchi, Huei-Ping Huang, et al. "Model Projections of an Imminent Transition to a More Arid Climate in Southwestern North America." *Science* 316 (May 25, 2007): 1181–1183.

Seck, Emmanuel. "National Action Programs for Climate Adaptation: A Déjà Vu or a Real Chance to Build on Past Experiences? A Presentation to the ACP-EU Council of Ministers—Specialist Conference Report on Governance and Combating Desertification." Brussels, Friedrich Ebert Stiftung, May 23, 2007.

Sen, Amartya. *Development as Freedom*. New York: Oxford University Press, 1999.

Senate Select Committee on Intelligence. *Annual Threat Assessment of the Intelligence Community*. 2009.

Shacknove, Andrew. "Who Is a Refugee?" *Ethics* 95: 2 (1985): 274–284.

Sindico, Francesco. "Climate Change: A Security (Council) Issue?" *Carbon and Climate Law Review* 1 (2007): 26–31.

Skeldon, Ronald. "Background Paper for Roundtable Session 3.2: Assessing the Relevance and Impact of Climate Change on Migration and Development." Global Forum on Migration & Development, Mexico, November 2010, available at www.gfmd.org.

Slaughter, Ann-Marie. "The Real New World Order." *Foreign Affairs* 183 (September/October 1997).

Slyomovics, Susan. *The Performance of Human Rights in Morocco*. Philadelphia: University of Pennsylvania Press, 2005.

Smith, Paul J. "Climate Change, Mass Migration and the Military Response." *Orbis* 51: 4 (Fall 2007): 617–633.

Solana, Javier, and the European Commission. *Climate Change and International Security: Paper from the High Representative and the European Commission to the European Council*. Brussels, Belgium: Council of the European Union, 2008.

Solé, Carlota, and Sònia Parella. "The Labour Market and Racial Discrimination in Spain." *Journal of Ethnic and Migration Studies* 29: 1 (2003): 121–140.

Sparke, Matthew. "Political Geography: Political Geographies of Globalization (1)—Dominance." *Progress in Human Geography* 28: 6 (2004): 777–794.

"Special Issue: Climate Change and Displacement." *Forced Migration Review* 31 (October 2008).

Stal, Marc. "Mozambique." In *EACH-FOR: Environmental Change and Forced Migration Scenarios D.3.4. Synthesis Report*, edited by Andras Vag. Brussels, Belgium: European Commission, 2009, 40–41.

Steiner, Achim. "Environment as a Peace Policy." *NATO Review: How Does NATO Need to Change (Parts 1 and 2)?*, available at www.nato.int/docu/review/2009/NATO_Change/Environment_PeacePolicy/EN/.

Stiglitz, Joseph. *Globalization and Its Discontents*. New York: Penguin Putnam, 2002.

Strondl, Robert. *Frontex: General Report for 2008*. Warsaw, Poland: Frontex, available at www.frontex.europa.eu, 2008.

Swearingen, Will. *Moroccan Mirages: Agrarian Dreams and Deceptions, 1912–1986*. Princeton, NJ: Princeton University Press, 1987.

Tarhule, Aondover, and Peter J. Lamb. "Climate Research and Seasonal Forecasting for West Africans." *Bulletin of the American Meteorological Society* 84 (2003): 1741–1759.

Terry, Geraldine. "No Climate Justice without Gender Justice: An Overview of the Issues." *Gender & Development* 17: 1 (March 2009): 5–18.

Tickner, Jill. *Gender in International Relations*. New York: Columbia University Press, 1992.

Todaro, Michael. *International Migration in Developing Countries: A Review of Theory.* Geneva, Switzerland: International Labour Organization, 1976.

Tonah, Steve. "Integration or Exclusion of Fulbe Pastoralists in West Africa: A Comparative Analysis of Interethnic Relations, State and Local Policies in Ghana and Côte d'Ivoire." *Journal of Modern African Studies* 41: 1 (2003): 91–114.

Toulmin, Camilla. *Climate Change in Africa.* London: Zed Books, 2010.

Trenberth, Kevin E., Aiguo Dai, Roy M. Rasmussen, and David B. Parsons. "The Changing Character of Precipitation." *Bulletin of the American Meteorological Society* 84 (September 2003): 1205–1217.

Trenberth, Kevin E., Philip D. Jones, Peter Ambenje, Roxana Bjariu, David Easterling, Albert Klein Tank, David Parker, et al. "Observations: Surface and Atmospheric Climate Change." In *Climate Change 2007: The Physical Science Basis. Contribution of Working Group I to the Fourth Assessment Report of the Intergovernmental Panel on Climate Change,* edited by S. Solomon, D. Qin, M. Manning, Z. Chen, M. Marquis, K. B. Averyt, M. Tignor, and H. L. Miller. New York: Cambridge University Press, 2007.

Unander, Fidtjof. *From Oil Crisis to Climate Challenge: Understanding CO2 Emission Trends in IEA Countries.* Paris, OECD and IEA, 2003.

United Kingdom Ministry of Defence's Development, Concepts and Doctrine Centre. *DCDC Global Strategic Trends Programme: 2007–2036.* 3rd ed. London: UK Ministry of Defence, 2007.

United Nations World Commission on Environment and Development. *Our Common Future.* New York: United Nations, 1987, available at www.un-documents.net/ocf-ov.htm.

Valenzuela, J. Samuel, and Arturo Valenzuela. "Modernization and Dependency: Alternative Perspectives in the Study of Latin American Underdevelopment." *Comparative Politics* 10: 4 (1978): 535–557.

Valluy, Jérôme. "Aux marches de l'Europe: des 'pay-camps': La transformation des pays de transit en pays d'immigration forcée (observations à partir de l'exemple marocain)." In *Immigration sur emigration: Le Maghreb à l'épreuve des migrations subsahariennes,* edited by Ali Bensaâd. Paris: Karthala, 2008, 325–342.

van der Geest, Kees. "Ghana." In *EACH-FOR: Environmental Change and Forced Migration Scenarios D.3.4. Synthesis Report,* edited by Andras Vag. Brussels, Belgium: European Commission, 2009, 46–47.

Vanderheiden, Steve. *Atmospheric Justice: A Political Theory of Climate Change.* New York: Oxford University Press, 2008.

Vecchi, Gabriel A., and Brian J. Soden. "Global Warming and the Weakening of the Tropical Circulation." *Journal of Climate* 20 (September 1, 2007): 4316–4340.

Vörösmarty, Charles J., Pamela Green, Joseph Salisbury, and Richard Lammers. "Global Water Resources: Vulnerability from Climate Change and Population Growth." *Science* 289 (July 14, 2000): 284–288.

Wæver, Ole. "Societal Security: The Concept." In *Identity, Migration and the New Security Agenda in Europe,* edited by Ole Wæver, Barry Buzan, Morten Kelstrup, and Pierre Lemaitre. New York: St. Martin's, 1993, 17–40.

Walker, R. B. J. "State Sovereignty and the Articulation of Political Space/Time." *Millennium* 20: 3 (1991): 445–49.

Wallerstein, Immanuel. *The Capitalist World Economy.* London: Cambridge University Press, 1979.

Walt, Stephen. "The Renaissance of Security Studies." *International Studies Quarterly* 35: 2 (1991): 211–239.

Walters, William. "Mapping Schengenland: Denaturalizing the Border." *Environment and Planning D: Society and Space* 20 (2002): 561–580.

Waltz, Susan. "The Politics of Human Rights in the Maghreb." In *Islam, Democracy, and the State in North Africa*, edited by John Entelis. Bloomington: Indiana University Press, 1997, 75–92.

Warner, Koko, Charles Ehrhart, Alex de Sherbinin, and Susana Adamo. *In Search of Shelter: Mapping the Effects of Climate Change on Human Migration and Displacement.* New York: CARE International, 2009.

Warren, A. "The Policy Implications of Sahelian Change." *Journal of Arid Environments* 63 (2005): 660–670.

Washington, Richard, Mike Harrison, Declan Conway, Emily Black, Andrew Challinor, David Grimes, Richard Jones, Andy Morse, Gillian Kay, and Martin Todd. "African Climate Change: Taking the Shorter Route." *Bulletin of the American Meteorological Society* 87: 10 (2006): 1355–1366.

Waterbury, John. *The Commander of the Faithful: The Moroccan Political Elite—A Study in Segmented Politics.* New York: Columbia University Press, 1970.

Weiner, Myron. *Global Migration Crisis: Challenges to States and to Human Rights.* New York: Harper Collins, 1995.

Wendt, Alexander. "Anarchy Is What States Make of It." *International Organization* 42: 2 (1992): 391–425.

Weston, Burns, and Tracy Bach. *Recalibrating the Law of Humans with the Laws of Nature: Climate Change, Human Rights, and Intergenerational Justice*: Vermont Law School Legal Studies Research Paper Series No. 10-06, 2009.

Wheeler, Nicholas. *Saving Strangers: Humanitarian Intervention in International Society.* New York: Oxford University Press, 2003.

White, Gregory. " 'The End of the Era of Leniency' in Morocco? Mohammed VI's Halting Glasnost." In *North Africa: Politics, Religion and the Limits of Transformation*, edited by Y. Zoubir and H. Amirah-Fernández. London: Routledge, 2008, 90–108.

———. "Free Trade as a Strategic Instrument in the War on Terror? The 2004 U.S.-Moroccan Free Trade Agreement." *Middle East Journal* 59: 4 (Fall 2005), 597–616.

———. "It's Not Because They're Arab," *Open Democracy*, February 10, 2011, available at www.opendemocracy.net/gregory-white/it's-not-because-they're-arab.

———. "The Maghreb in the World's Political Economy." *Middle East Policy* 14: 4 (Winter 2007): 42–54.

———. "La migración laboral Marroquí y los territorios Españoles de Ceuta y Melilla." *Revista Internacional De Sociología* 36 (September–December 2003): 135–168.

———. *On the Outside of Europe Looking In: A Comparative Political Economy of Tunisia and Morocco.* Albany, NY: State University of New York Press, 2001.

———. "Sovereignty and International Labor Migration: The 'Security Mentality' in Spanish-Moroccan Relations as an Assertion of Sovereignty." *Review of International Political Economy* 14: 4 (2007): 690–718.

———. "Too Many Boats, Not Enough Fish: The Political Economy of Morocco's 1995 Fishing Accord with the European Union." *Journal of Developing Areas* 31: 3 (1997): 313–336.

Williams, John. *The Ethics of Territorial Borders: Drawing Lines in the Shifting Sands.* London: Palgrave, 2006.

Williams, Michael C. "Words, Images, Enemies: Securitization and International Politics." *International Studies Quarterly* 47 (2003): 511–531.

Witze, Alexandra. "Losing Greenland." *Nature* 452 (April 17, 2008): 798–802.

Young, Oran. *International Governance: Protecting the Environment in a Stateless Society*. Ithaca, CA: Cornell University Press, 1994.

Zenawi, Meles. *Statement on Behalf of the African Group at the COP 15*. Copenhagen, Denmark, 2009.

Zetter, Roger. "Legal and Normative Frameworks." *Forced Migration Review* 31 (2008): 62–63.

———. "The Role of Legal and Normative Frameworks for the Protection of Environmentally Displaced People." In *Migration, Environment and Climate Change: Assessing the Evidence*, edited by Frank Laczko and Christine Aghazarm. Geneva, International Organization for Migration, 2010, 385–441.

Zolberg, Aristide R. "Beyond the Crisis." In *Global Migrants, Global Refugees: Problems and Solutions*, edited by Aristide Zolberg and Peter Benda. New York: Berghahn, 2001, 1–16.

———. "Patterns of International Migration Policy: A Diachronic Comparison." In *Minorities: Community and Identity*, edited by Charles Fried. Berlin: Springer-Verlag, 1983, 229–246.

INDEX

England, 67, 70
   See also European Union (EU);
      United Kingdom
ENSO (El Niño-Southern Oscillation).
   See El Niño-Southern Oscillation
      (ENSO)
Environmental Change and Security
   Program (Woodrow Wilson
   Center), 139
environmental migrants
   IOM definition of, 29, 59
   See also climate-induced migration
      (CIM)
environmental refugees
   as alternate term for CIM, 20, 131–132
   description of, 3, 21
   typology of, *26*
   See also climate-induced migration (CIM)
environmental security
   increasing attention to, 3–4, 62–65
   as rationale for anti-immigration
      stance, 5–6, 11, 68–69, 88
   See also securitization of
      climate-induced migration
equatorial regions, 40–41
ethics
   of border security, 6–7, 16, 126–127,
      128
   of climate-induced migration, 6–7,
      13, 20, 128, 133
   and international relations, 127–128
   "lifeboat," 50–51
   and responsibility for climate change,
      19, 128
   and sea-level rise, 133
EU-Morocco Summit, 119–120
Euro-Mediterranean Partnership, 66
Europe
   coal use, 137
   immigration dynamics in, 14–15,
      98–101
   securitization of CIM, 66–67, 68,
      75–78, 146
   See also African migration to Europe;
      European Union (EU)
European Commission, 76, 77–78, 105,
   119
European Council, 83, 119
European Economic Community (EEC),
   14, 97, 104

European Maritime Safety Agency
   (EMSA), 76
European Police College (CEPOL), 76
European Union (EU)
   Africa-EU Summits, 92, 118–119
   Common Agricultural Policy (CAP),
      102
   energy policy, 56
   establishment of, 66
   joint efforts with Morocco, 113,
      117–118, 119–120
   and Kyoto Protocol, 56
   migration control through transit
      states, 7–9, 76–77, 90
   securitization of immigration,
      14–15, 66–67, 75–76, 97
   See also Frontex; North Atlantic
      countries; Schengen Agreement;
      Schengen Area
Europe Partner States (EPS), 117
Europol, 76, 105

Falk, Richard, 63
Fatal Climate (Hood), 135n30
Ferguson, Niall, 142
Findley, Sally, 49
Fingar, Thomas, 79
Flatow, Ira, xi
flooding, 26, 27, 56
   See also sea-level rise
Food and Agricultural Organization
   (FAO), 131
Fourier, Joseph, 34
"fragmented regime complex," 130–131
France
   immigration dynamics, 66, 70, 76,
      98, 99
   terrorist attack, 67, 124
French Development Agency (AFD), 85
Friends and Families of Victims of
   Clandestine Immigration (AFVIC),
   110
Frontex
   establishment of, 14–15, 76
   map of migration routes into
      Morocco, 105, 106, *107*
   mission of, 15, 76
   and NATO, 82, 83
Funk, Chris, 39, 41
Furth, Leon, 86

tendency toward short-distances in, 47–49

varied reasons for, 47

See also climate-induced migration (CIM); immigration; migrants

Milankovitch, Milutin, 35

military groups
 and securitization of CIM, 57, 63, 83–84
 See also U.S. military

Millennium Development Goals (MDGs), 54, 140–141

Millennium Villages project, 140

Ministry of the Moroccan Community Abroad (MMCA), 111

mitigation of greenhouse gases
 in Africa, 45
 assessing costs of, 80–81, 88
 distractions from, 125–126
 ethics of, 128
 by G20 countries, 136, 142–143
 political unpopularity of, 7, 58
 urgent need for, 7, 136–137, 144
 See also CO_2 emissions

Mohammed V Foundation for Solidarity, 111

Mohammed VI, 8, 92, 113, 118, 147

Montreal Protocol on Substances That Deplete the Ozone Layer, 131

Morisetti, Neil, 145

Moroccan Association for the Research and Study of Migration (AMERM), 104–105

Moroccan Organization for Human Rights (OMDH), 110

Morocco
 agricultural production, 101–102
 counterterrorism efforts, 97
 ethnic tensions, 101
 French and Spanish colonialism, 98, 104
 human rights record, 114, 122, 147
 reign of Hassan II, 112, 147
 reign of Mohammed VI, 8, 92, 113, 118, 147
 relative stability, 147
 and Strait of Gibraltar, 17
 territorial claims over Western Sahara, 16, 97, 120, 122
 terror attacks in, 114

as transit state, 7, 11, 25, 91, 95, 97, 120, 147

2009 IUSSP meeting, 92

vulnerability to climate change, 120

See also Ceuta and Melilla; Morocco emigration/immigration dynamics

Morocco emigration/immigration dynamics
 artistic treatment of, 103
 CIM as rationale for, 118–120
 emigration dynamics, 98–103, 110–111, 115, 120
 immigration dynamics, 8, 103–111
 impact on governance and diplomacy, 120, 121–123
 joint efforts with EU, 113, 117–118, 119–120
 joint efforts with NATO, 82, 83–84, 97, 109, 113, 117
 joint efforts with Spain, 115–117
 legal vs. illegal migration, 102–103
 passage of migration Law 02-03, 113–115
 remittances and social capital of migrants, 102
 routes for, 105–106, *107*
 state response to, 109–115
 and terrorism concerns, 113, 114, 117
 transit migration control in, 8, 111–115, 116, 119, 120
 treatment of undocumented migrants, 115, 116
 See also Ceuta and Melilla

Mortimore, Michael, 43–44

Mozambique, 54, 74

Mubarak, Hosni, 145

multilateralism
 as climate change strategy, 80, 142–143
 in global fiscal crisis, 142

Munich Re Foundation, 132

Myanmar (Burma), 56, 72

Myers, Norman, 3, 28, 69

NAFTA (North American Free Trade Agreement), 8, 67, 71, 97

NAPAs (National Adaptation Programmes of Action), 132, 141–142

Napolitano, Janet, 62